A HANDBOOK
OF ACTIVE FILTERS

D.E. JOHNSON, J.R. JOHNSON, and **H.P. MOORE**

*Department of Electrical Engineering
Louisiana State University*

PRENTICE-HALL, INC., *Englewood Cliffs, New Jersey* 07632

Library of Congress Cataloging in Publication Data

Johnson, David E
  A handbook of active filters.

  Bibliography: p.
  Includes index.
  1. Electric filters, Active. I. Johnson, Johnny Ray, joint author. II. Moore, Harry P., joint author. III. Title.
TK7872.F5J65   621.3815'32   79-10373
ISBN 0-13-372409-3

Editorial production supervision
and interior design by: JAMES M. CHEGE

Manufacturing buyer: GORDON OSBOURNE

© 1980 by Prentice-Hall, Inc., Englewood Cliffs, N.J. 07632

All rights reserved. No part of this book may be reproduced in any form or by any means without permission in writing from the publisher.

Printed in the United States of America

10 9 8 7 6 5 4 3 2 1

PRENTICE-HALL INTERNATIONAL, INC., *London*
PRENTICE-HALL OF AUSTRALIA PTY. LIMITED, *Sydney*
PRENTICE-HALL OF CANADA, LTD., *Toronto*
PRENTICE-HALL OF INDIA PRIVATE LIMITED, *New Delhi*
PRENTICE-HALL OF JAPAN, INC., *Tokyo*
PRENTICE-HALL OF SOUTHEAST ASIA PTE. LTD., *Singapore*
WHITEHALL BOOKS LIMITED, *Wellington, New Zealand*

To the memory of
Al Eskandar
A prince of a human being

CONTENTS

PREFACE   xi

1 INTRODUCTION   1

    1-1 Frequency-Selective Filters   *1*
    1-2 Other Types of Filters   *4*
    1-3 Transfer Functions   *5*
    1-4 Active Filter Elements   *6*
    1-5 Construction of Filters   *8*

2 LOW-PASS FILTERS—BUTTERWORTH AND CHEBYSHEV TYPES   11

    2-1 The General Low-Pass Case   *11*
    2-2 Butterworth Filters   *14*

vi   CONTENTS

- 2-3 Chebyshev Filters   *17*
- 2-4 Selection of the Minimum Order   *20*
- 2-5 Infinite-Gain Multiple-Feedback Low-Pass Filters   *22*
- 2-6 VCVS Low-Pass Filters   *26*
- 2-7 Biquad Low-Pass Filters   *28*
- 2-8 Tuning the Second-Order Filters   *29*
- 2-9 Odd-Order Filters   *31*
- 2-10 MFB Low-Pass Filter Design Summary   *33*
- 2-11 VCVS Low-Pass Filter Design Summary   *34*
- 2-12 Biquad Low-Pass Filter Design Summary   *36*
- 2-13 Odd-Order Low-Pass Filter Design Summary   *38*

3 LOW-PASS FILTERS—INVERSE CHEBYSHEV AND ELLIPTIC TYPES   41

- 3-1 Inverse Chebyshev Filters   *41*
- 3-2 Elliptic Filters   *46*
- 3-3 VCVS Elliptic Filter Circuits   *50*
- 3-4 Elliptic Filter Circuits with Three Capacitors   *54*
- 3-5 Biquad Elliptic Filter Circuits   *57*
- 3-6 Tuning the Inverse Chebyshev and Elliptic Filters   *58*
- 3-7 Odd-Order Elliptic Filters   *60*
- 3-8 VCVS Low-Pass Elliptic Filter Design Summary   *61*
- 3-9 Three-Capacitor Low-Pass Elliptic Filter Design Summary   *63*
- 3-10 Biquad Low-Pass Elliptic Filter Design Summary   *65*
- 3-11 Odd-Order Low-Pass Elliptic Filter Design Summary   *66*

4 HIGH-PASS FILTERS 69

4-1 The General Case 69
4-2 Infinite-Gain Multiple-Feedback High-Pass Filters 72
4-3 VCVS High-Pass Filters 74
4-4 Biquad High-Pass Filters 75
4-5 Inverse Chebyshev and High-Pass Elliptic Filter Circuits 76
4-6 Tuning the Second-Order Filters 77
4-7 Odd-Order High-Pass Filters 79
4-8 MFB High-Pass Filter Design Summary 80
4-9 VCVS High-Pass Filter Design Summary 81
4-10 Biquad High-Pass Filter Design Summary 83
4-11 VCVS High-Pass Elliptic Filter Design Summary 85
4-12 Three-Capacitor High-Pass Elliptic Filter Design Summary 87
4-13 Biquad High-Pass Elliptic Filter Design Summary 88
4-14 Odd-Order High-Pass Elliptic Filter Design Summary 90

5 BANDPASS FILTERS 93

5-1 The General Case 93
5-2 Transfer Functions 98
5-3 Transition Widths 101
5-4 Infinite-Gain Multiple-Feedback Bandpass Filters 103
5-5 VCVS Bandpass Filters 105
5-6 Biquad Bandpass Filters 106

viii CONTENTS

- 5-7 Inverse Chebyshev and Elliptic Bandpass Filters *107*
- 5-8 Tuning the Second-Order Bandpass Stages *108*
- 5-9 General Design Information for Bandpass Filter Construction *110*
- 5-10 MFB Bandpass Filter Design Summary *111*
- 5-11 VCVS Bandpass Filter Design Summary *114*
- 5-12 Biquad Bandpass Filter Design Summary *117*
- 5-13 VCVS Bandpass Elliptic Filter Design Summary *119*
- 5-14 Three-Capacitor Bandpass Elliptic Filter Design Summary *123*
- 5-15 Biquad Bandpass Elliptic Filter Design Summary *125*

6 BAND-REJECT FILTERS   129

- 6-1 The General Case *129*
- 6-2 Transfer Functions *133*
- 6-3 Transition Widths *135*
- 6-4 Infinite-Gain Multiple-Feedback Band-Reject Filters *137*
- 6-5 VCVS Band-Reject Filters *138*
- 6-6 Tuning the Second-Order Band-Reject Stages *140*
- 6-7 General Design Information for Band-Reject Filter Construction *141*
- 6-8 MFB Band-Reject Filter Design Summary *142*
- 6-9 VCVS Band-Reject Filter Design Summary *144*
- 6-10 VCVS Band-Reject Elliptic Filter Design Summary *145*
- 6-11 Three-Capacitor Band-Reject Elliptic Filter Design Summary *149*
- 6-12 Biquad Band-Reject Elliptic Filter Design Summary *153*

CONTENTS

7 ALL-PASS AND CONSTANT-TIME-DELAY FILTERS 157

- 7-1 All-Pass Filters *157*
- 7-2 Multiple-Feedback All-Pass Filters *159*
- 7-3 Biquad All-Pass Filters *161*
- 7-4 Bessel Filters *163*
- 7-5 All-Pass Constant-Time-Delay Filters *166*
- 7-6 MFB All-Pass Filter Design Summary *168*
- 7-7 Biquad All-Pass Filter Design Summary *170*
- 7-8 Bessel (Constant-Time-Delay) Filter Design Summary *172*
- 7-9 All-Pass Constant-Time-Delay Filter Design Summary *173*

APPENDIXES 175

- A Butterworth and Chebyshev Low-Pass Filter Data *177*
- B Inverse Chebyshev Low-Pass Filter Data *180*
- C Elliptic Low-Pass Filter Data *190*
- D Transition Widths for Elliptic Filters *231*
- E Bessel Low-Pass Filter Data *234*

REFERENCES 235

INDEX 239

PREFACE

In this book simplified, rapid methods are presented for obtaining complete, practical active filter designs by substituting numbers into equations. The book should be useful to all filter designers from the novice to the expert, since the tedious work of obtaining the design formulas has already been done and the filter characteristics have been tabulated. The circuit elements used are integrated-circuit operational amplifiers, resistances, and capacitances. The design formulas yield standard, commonly available element values.

A distinguishing feature of the book is that, in addition to the Butterworth and Chebyshev characteristics usually tabulated in books of this type, inverse Chebyshev and elliptic filter characteristics are included. The orders given in all four cases are 2 through 10. Also, Bessel filter characteristics of orders 2 through 6 are included.

From the design formulas one may construct the following

types of filters:

1. Butterworth, Chebyshev, inverse Chebyshev, and elliptic filters of low-pass and high-pass types, of orders ranging from 2 to 10, and bandpass and band-reject types of orders 2, 4, 6, ..., 20
2. Phase-shift or all-pass filters of order 2
3. Constant-time-delay or Bessel filters of orders 2 through 6
4. All-pass, constant-time-delay filters of orders 2 through 6

In the Chebyshev and elliptic cases, passband ripple widths of 0.1, 0.5, 1, 2, and 3 dB are available, and in the inverse Chebyshev and elliptic cases, minimum stopband loss values are tabulated from 30 to 100 dB, or so, in steps of 5 dB.

The most commonly used filter circuits are presented for each filter type, ranging in complexity from voltage-controlled-voltage-source types with one op amp to biquad circuits with three op amps.

Each filter type is discussed in a separate chapter, and summaries of each procedure are given at the end of the chapter, where practical design suggestions are included. Numerous examples are given in detail for most of the filter types, and photographs of the actual filter amplitude responses are shown. Tuning procedures are given for each type and transition widths are either tabulated (in the elliptic case) or are given by formulas.

A very desirable feature of the book is that each chapter is independent of the others and the design summaries at the end of the chapter are independent of the rest of the chapter. Thus a novice or an expert may use the summaries to design practical filters without reading or understanding the chapters.

There are many people who have provided invaluable assistance in the development of this book. In particular, a special note of thanks is due Mrs. Marie Jines and Mrs. Norma Duffy for the expert typing and draftsmanship, respectively.

DAVID E. JOHNSON
JOHNNY R. JOHNSON
HARRY P. MOORE

Louisiana State University
Baton Rouge, LA.

1
INTRODUCTION

1-1 FREQUENCY-SELECTIVE FILTERS

An electric filter is, in most cases, a *frequency-selective* device. That is, it passes signals of certain frequencies and blocks or attenuates signals of other frequencies. The most common types of frequency-selective filters are *low-pass* (which pass low frequencies and block high frequencies), *high-pass* (which pass high frequencies and block low frequencies), *bandpass* (which pass a *band* of frequencies and block those frequencies that are higher and lower than those in the band), and *band-reject filters* (which block a band of frequencies and pass those frequencies that are higher and lower than those in the band).

The performance of a frequency-selective filter may be described more precisely by a consideration of its transfer function $H(s)$, which we take as

$$H(s) = \frac{V_2(s)}{V_1(s)} \qquad (1\text{-}1)$$

The quantities V_1 and V_2 are, respectively, the input and output voltages, as shown in the general representation of Fig. 1-1. At the steady-state frequency $s = j\omega$ ($j = \sqrt{-1}$), the transfer function

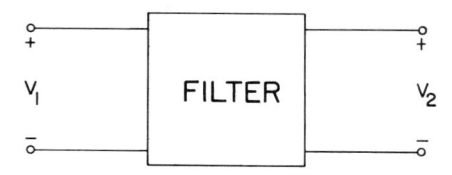

FIGURE 1-1 An electric filter representation.

may be written in the form

$$H(j\omega) = |H(j\omega)| e^{j\phi(\omega)} \tag{1-2}$$

where $|H(j\omega)|$ is the *amplitude* or *magnitude*, $\phi(\omega)$ is the *phase*, and ω (radians/second) is related to the frequency f (hertz) by $\omega = 2\pi f$.

The frequencies that pass are in certain ranges, or bands, called *passbands*, in which the amplitude $|H(j\omega)|$ is relatively large and ideally is constant. The bands of frequencies that are blocked are *stopbands*, in which the amplitude is relatively small and ideally is zero. As an example, the amplitude shown by the dashed line of Fig. 1-2 is that of an *ideal* low-pass filter with a single passband $0 < \omega < \omega_c$ and stopband $\omega > \omega_c$. The frequency ω_c (or in hertz, $f_c = \omega_c/2\pi$) between the two bands is the *cutoff frequency*.

In actual practice it is impossible to achieve the ideal response because of the sharp corners it requires. A central problem in filter design is, therefore, to approximate the ideal response to some prescribed degree of accuracy with a practical response, which can be obtained in the laboratory. One such practical response is represented by the solid line in Fig. 1-2.

In the practical case the passband and stopband are not clearly demarcated and must be formally defined. As our definition we take the passband as the range of frequencies where the amplitude is greater than some specified number, shown as A_1 in Fig. 1-2, and the stopband as the range where the amplitude is less than some specified amount, such as A_2. The interval in which the response continually decreases from the passband to the stopband is called the *transition band*. The practical example of Fig. 1-2 has

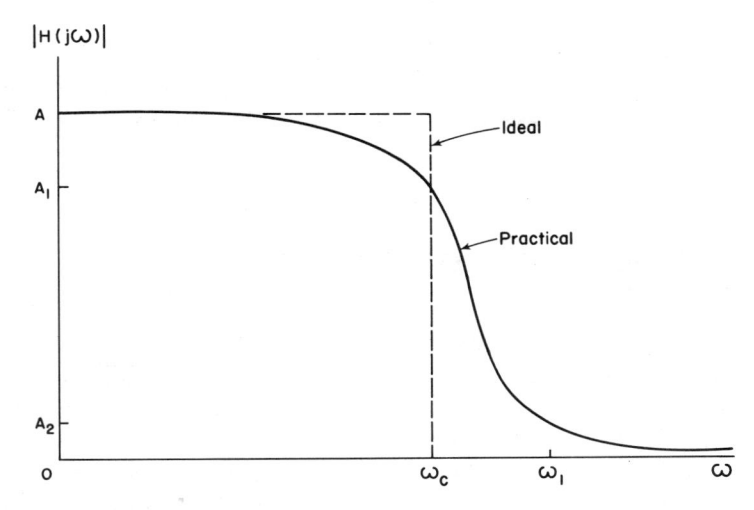

FIGURE 1-2 *Ideal and practical low-pass filter amplitudes.*

passband $0 < \omega < \omega_c$, stopband $\omega > \omega_1$, and transition band $\omega_c < \omega < \omega_1$.

The amplitude may also be expressed in *decibels* (dB), defined by

$$\alpha = -20 \log_{10} |H(j\omega)| \quad \text{dB} \qquad (1\text{-}3)$$

in which case α is referred to as *loss*. For example, suppose that we take $A = 1$ in Fig. 1-2, corresponding to $\alpha = 0$ dB. Then if $A_1 = A/\sqrt{2} = 1/\sqrt{2}$, the loss at ω_c is

$$\alpha_1 = -20 \log_{10}\left(\frac{1}{\sqrt{2}}\right)$$
$$= 10 \log 2$$
$$= 3 \text{ dB}$$

Generally, the loss in the passband never exceeds 3 dB; thus from this example we see that the passband amplitudes are at least $1/\sqrt{2} = 0.707$, or 70.7% of the maximum amplitude. In this case we could say also that in the passband the amplitude is *down* 3 dB or less from its maximum value.

In the case of frequency-selective filters, the amplitude is the most important characteristic, since its value at a certain frequency determines whether that frequency passes or is blocked. In this book we shall be primarily interested in frequency-selective filters, but we shall also consider two other types of filters—the all-pass filter and the constant-time-delay filter. In the next section we shall consider briefly other transfer-function characteristics that are important to these filter types.

1-2 OTHER TYPES OF FILTERS

In addition to frequency-selective filters, we may have filters in which the phase response $\phi(\omega)$ of Eq. (1-2) is the important characteristic. For example, an *all-pass filter* is one for which the amplitude is constant for all frequencies (all frequencies pass equally well) and the phase response is a function of frequency. An all-pass filter is thus a *phase-shifting filter*, since its amplitude is unchanged while its phase may be varied or shifted by changing the frequency.

In general, phase is important, though not the overriding characteristic of interest, in frequency-selective filters. This is because the output voltage is an *undistorted* version of the input voltage if it is the input voltage *amplified* and/or *delayed* in time. In this case the amplitude response is constant and the phase response is *linear*, given by

$$\phi(\omega) = -\tau\omega \qquad (1\text{-}4)$$

where τ is a constant (see, for example, [32, 33, 5, 16]*). The more nonlinear the phase response, the more distorted will be the output signal. Unfortunately, as the amplitude response improves (approaches the ideal case), the phase response deteriorates, and vice versa. Thus filter design involves a compromise between good amplitude and good phase response.

The *time delay* $T(\omega)$ of a filter is defined to be the negative of the slope of the phase response. That is,

$$T(\omega) = -\frac{d}{d\omega}\phi(\omega) \qquad (1\text{-}5)$$

Thus in the case of linear phase we have, from Eq. (1-4), $T(\omega) =$

---

* References thus cited are listed in numerical order in the References at the end of the book.

τ, a constant. A *time-delay filter* is one in which the time delay is the overriding characteristic of interest, and is designed so that $T(\omega)$ is very nearly constant over a specified frequency range.

In this book we shall be interested primarily in frequency-selective filters, but we shall also give some consideration to other types. In any case it is well to keep in mind the properties of phase shift and time delay, important in these other filter types. The distortion in the output signal is related directly to phase, which is related to time delay.

1-3 TRANSFER FUNCTIONS

As was stated earlier, ideal filters are impossible to construct, but approximations to ideal behavior may be obtained with practical, or *realizable*, filters (those that can be constructed, or realized, with actual hardware). The transfer function of a realizable filter is a ratio of polynomials, which for our purpose we shall take as

$$H(s) = \frac{V_2(s)}{V_1(s)} = \frac{a_m s^m + a_{m-1} s^{m-1} + \cdots + a_1 s + a_0}{b_n s^n + b_{n-1} s^{n-1} + \cdots + b_1 s + b_0} \quad (1\text{-}6)$$

The a's and b's are real constants and

$$m, n = 1, 2, 3, \ldots \quad (m \leq n) \quad (1\text{-}7)$$

The degree n of the denominator polynomial is defined to be the *order* of the filter. As we shall see, practical amplitudes are better (more nearly ideal) for higher-order than for lower-order filters. However, higher order is accompanied by more complex circuits and higher cost. One aspect of filter design is thus to obtain a realizable response that approximates to some prescribed degree of accuracy the ideal response and is also as economical as possible.

If all the a's in Eq. (1-6) are zero except a_0, the transfer function is the ratio of a constant to a polynomial. In this case the filter is called an *all-pole filter*, because its transfer function has the property that all its poles and none of its zeros are finite. (A zero is a value of s at which the transfer function is zero and a pole is a value of s at which the transfer function becomes infinite.)

In succeeding chapters we shall consider shortcut methods of obtaining practical filters of various types. We shall consider cases based on all-pole as well as on more general transfer functions.

1-4 ACTIVE FILTER ELEMENTS

Once a suitable transfer function is obtained, the construction of the filter is completed by obtaining a circuit with the given transfer function. This may be done in a variety of ways, all of which may be broadly classified as either passive or active methods.

Passive filters are those constructed with resistors, capacitors, and inductors, all of which are passive circuit elements. These filters are very useful for operation in certain frequency ranges but are quite undesirable at low frequencies, say 0.5 MHz or less. This is because at low frequencies the inductors required are unsatisfactory because of their size and considerable departure from their ideal behavior. Also, unlike resistors and capacitors, inductors cannot be readily adapted to integrated-circuit techniques.

Thus for low-frequency applications it is desirable to eliminate inductors from the filter circuits. This may be done by constructing *active filters* using resistors, capacitors, and one or more active devices, such as transistors, dependent sources, and so on (see, for example, [28, 21, 22, 13, 14]). One of the most often-used active devices [12, 17, 4, 9, 16, 6], and the one we shall use exclusively, is the integrated-circuit (IC) *operational amplifier*, or *op amp*, the symbol for which is shown in Fig. 1-3.

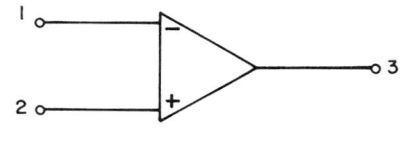

FIGURE 1-3 An op amp.

The op amp is a multiterminal device, but for simplicity we have shown only three terminals. These are the *inverting input* (1), the *noninverting input* (2), and the *output* (3) terminals. In the ideal case the op amp has infinite input resistance, zero output resistance, and infinite gain [9]. As a consequence, the circuit analyst may consider the voltage between the input terminals as well as the current into the input terminals to be zero. Practical op amps approximate the ideal very closely over a limited frequency range that varies with the type of op amp.

The terminals that are not shown in Fig. 1-3 are normally power supply terminals; compensation terminals required on op amps, such as the 709 type; and offset null terminals provided on those, such as the 741 type. These additional terminals are used in accordance with specifications supplied by the manufacturer. In general, externally compensated op amps give superior results at higher frequencies than do internally compensated op amps (those with no compensation terminals, such as the 741).

In constructing an active filter the designer should use op amps that are adequate for the specified gains and frequency ranges. For example, the open-loop gains of the op amps should be at least 50 times the filter gain [20]. (Later we shall define the term "filter gain," which varies with the type of filter under consideration.)

Also, for best operation, consideration should be given to the *slew rate* of the op amp. This is a number usually given in volts per microsecond, representing the limiting value of output voltage swing at a given frequency that can be accommodated by the op amp. For applications requiring large output voltage swings, op amps with high slew rates will be required. Slew rates typically vary from 0.5 to several hundred V/μs; however, some special-purpose op amps have slew rates of several thousand V/μs.

Information on open-loop gains, slew rates, terminal connections, and so on, is given in detail in the catalogs provided by manufacturers of op amps. In addition, there are many other publications with detailed sections devoted to op-amp characteristics (see, for example, [9, 20, 10]). Well-known manufacturers of op amps include Texas Instruments, Fairchild Semiconductor, Burr-Brown Research Corporation, National Semiconductor, Signetics Corporation, Motorola, and RCA.

The most commonly used and inexpensive resistors are the carbon composition type, which may be used in noncritical filter designs. In the case of fourth- or lower-order filters, 5% tolerance carbon composition resistors are often adequate, particularly if the filter is to perform at room temperature. For high-performance filters, higher-quality resistors, such as metal-film and wire-wound types, should be used. The higher the order, the lower the tolerances should be. Filters of orders higher than four should be constructed with resistors having tolerances of 2% or less.

In the case of capacitors, an acceptable common type is the Mylar capacitor, which can be used successfully in most filter designs. Polystyrene and Teflon capacitors are better, but more expensive choices for high-performance filters. The common, economical ceramic disk capacitor should be used only in the least critical applications.

1-5 CONSTRUCTION OF FILTERS

There are a number of ways of constructing a filter with a given nth-order transfer function. One popular method is to express the transfer function as a product of factors $H_1, H_2, \ldots, H_m$, and construct circuits, or *sections*, or *stages*, $N_1, N_2, \ldots, N_m$, corresponding to each factor. Finally, the stages are *cascaded* (the output of the first is the input of the second, and so forth) as shown in Fig. 1-4. If the stages do not interact with each other to change their

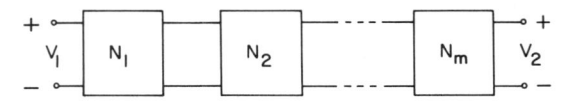

FIGURE 1-4 A cascading of stages.

individual transfer functions, the overall circuit has the given nth-order transfer function. The infinite input and zero output resistance characteristics of the op amps, referred to earlier, can be used to produce noninteracting stages.

In the case of first-order filters, the transfer function is of the form

$$\frac{V_2}{V_1} = \frac{P(s)}{s + C} \qquad (1\text{-}8)$$

where C is a constant and $P(s)$ is a polynomial of degree one or zero. A second-order filter has a transfer function of the form

$$\frac{V_2}{V_1} = \frac{P(s)}{s^2 + Bs + C} \qquad (1\text{-}9)$$

where B and C are constants and $P(s)$ is a polynomial of degree two or less.

For even order $n > 2$, the usual cascaded circuit has $n/2$ second-order stages, each with a transfer function like Eq. (1-9). If the order $n > 2$ is odd, there will be $(n-1)/2$ second-order stages with transfer functions like Eq. (1-9) and one first-order stage with a transfer function like Eq. (1-8).

In the case of Eq. (1-9), we define the *pole-pair frequency* by

$$\omega_p = \sqrt{C} \tag{1-10}$$

and the *pole-pair quality factor* by

$$Q_p = \frac{\sqrt{C}}{B} \tag{1-11}$$

Thus we may write Eq. (1-9) in the form

$$\frac{V_2}{V_1} = \frac{P(s)}{s^2 + (\omega_p/Q_p)s + \omega_p^2} \tag{1-12}$$

As we shall see, if Q_p is low, say between 0 and 5, relatively simple circuits may be used to achieve Eq. (1-9). However, for high Q_p, say 10 or more, more complex circuits may be required.

2

LOW-PASS FILTERS— BUTTERWORTH AND CHEBYSHEV TYPES

2-1 *THE GENERAL LOW-PASS CASE*

A *low-pass filter* is one that passes signals with low frequencies and blocks those with high frequencies. In the general case we shall take the passband as the interval $0 < \omega < \omega_c$, the stopband as $\omega > \omega_1$, the transition band as $\omega_c < \omega < \omega_1$, and, of course, ω_c as the cutoff frequency. These are shown with a practical low-pass amplitude in Fig. 2-1, where the shaded areas represent the allowable deviations in a given case in the pass- and stopbands.

If we normalize the minimum loss to 0 dB (for $A = 1$ in Fig. 2-1), the low-pass response in decibels is as shown in Fig. 2-2. The maximum passband loss is α_1 dB and the minimum stopband loss is α_2 dB (corresponding respectively to amplitudes of A_1 and A_2). The loss α_1 cannot exceed 3 dB, and typical values of α_2 are considerably larger, possibly in the range $20 \leq \alpha_2 \leq 100$ dB (in which case we have $0.1 \geq A_2 \geq 0.00001$).

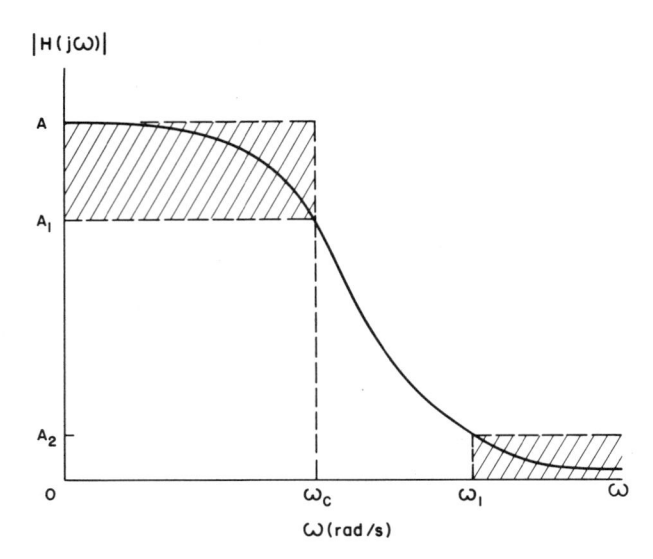

FIGURE 2-1 A typical low-pass amplitude.

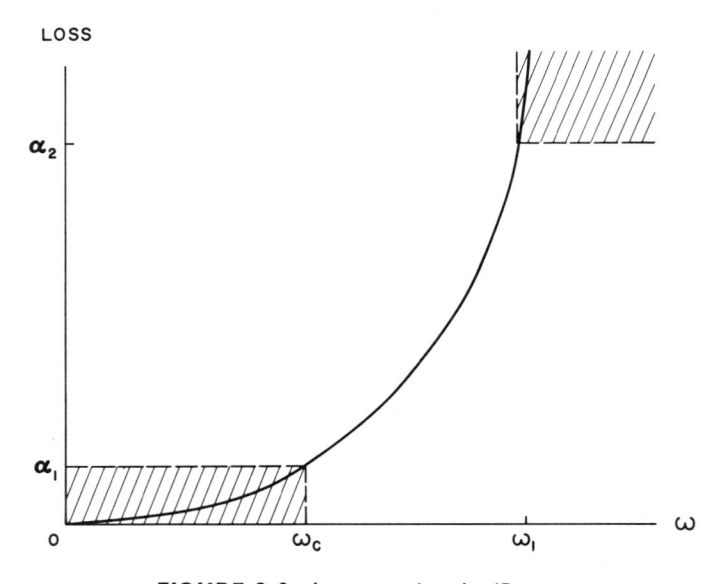

FIGURE 2-2 Low-pass loss in dB.

The *gain* of a low-pass filter is the value of its transfer function at $s = 0$ or, equivalently, the value of the amplitude at $\omega = 0$. Thus the gain of the practical filter with amplitude shown in Fig. 2-1 is A.

For a given set of specifications, such as A, A_1, A_2, ω_c, and

SEC. 2-1   THE GENERAL LOW-PASS CASE   13

ω_1 in Fig. 2-1 or α_1, α_2, ω_c, and ω_1 in Fig. 2-2, there are many types of low-pass filters. The four most popular types are the *Butterworth*, *Chebyshev*, *inverse Chebyshev*, and *elliptic filters*. The Butterworth filter has a monotonic response such as that of Figs. 2-1 and 2-2. (A response is monotonically decreasing if it never increases with frequency.) The Chebyshev response has passband ripples and is monotonic in the stopband. A sixth-order Chebyshev response is shown in Fig. 2-3. The inverse Chebyshev response is

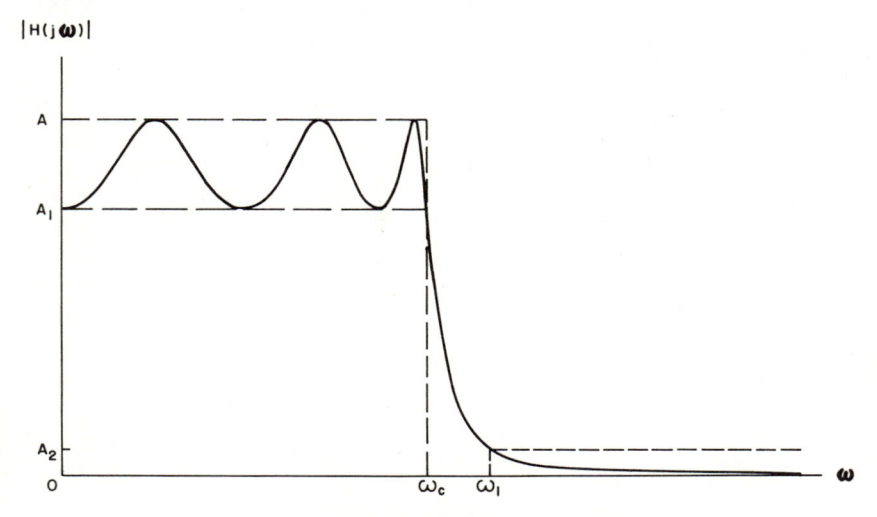

FIGURE 2-3 *A sixth-order Chebyshev response.*

monotonic in the passband and has stopband ripples, as shown in the sixth-order example of Fig. 2-4. Finally, the elliptic filter response has ripples in both the pass- and stopbands, as seen in the sixth-order example of Fig. 2-5.

The *optimum* low-pass filter is the one whose amplitude satisfies the conditions of Fig. 2-1 (or of Fig. 2-2) for a given order n and given allowable passband and stopband deviations, with a *minimum* transition width. That is, if A, A_1, A_2, n, and ω_c are given, then ω_1 is a minimum. In the all-pole case, the optimum filter is the Chebyshev [25]. However, in the general case, the elliptic filter is optimum [24] and is far superior to the Chebyshev filter.

In the sections to follow we shall briefly discuss the Butterworth and Chebyshev filters, which are the best known examples of all-pole filters. In Chapter 3 we shall consider the inverse Chebyshev and the elliptic filters, which have more general transfer

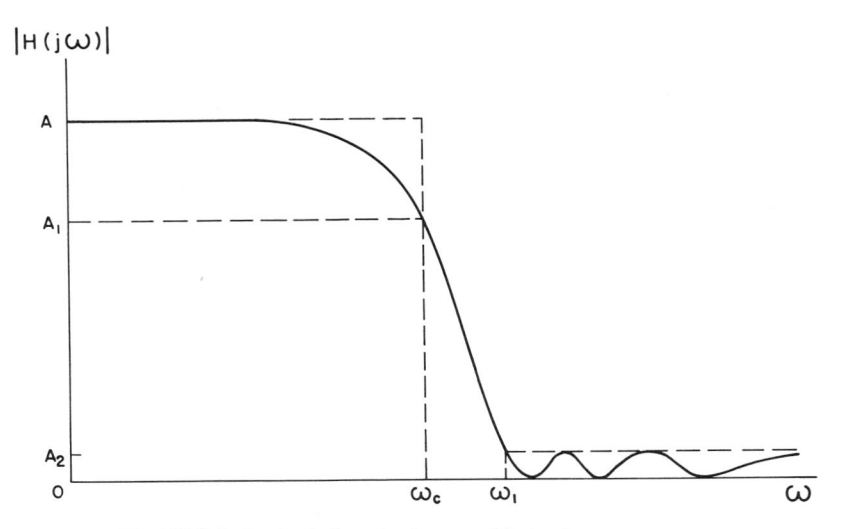

FIGURE 2-4 *A sixth-order inverse Chebyshev response.*

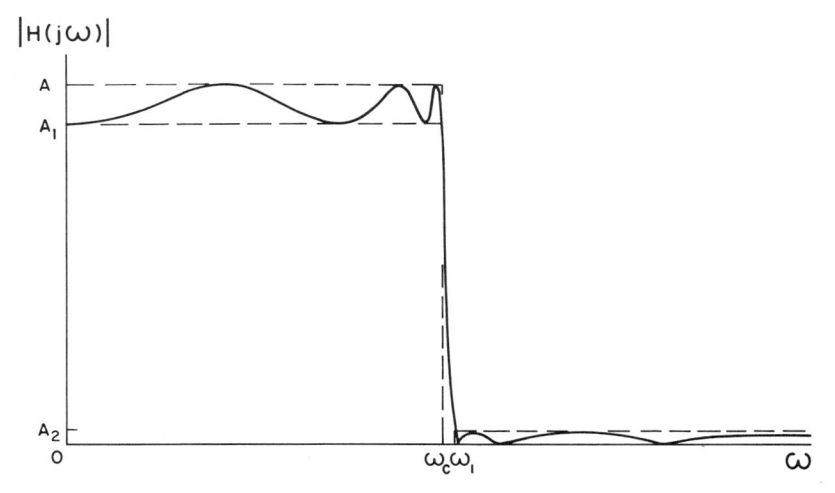

FIGURE 2-5 *A sixth-order elliptic response.*

functions. For a more detailed treatment, the reader is referred to such circuits or filter books as [32, 33, 5, 16, 29, 4, 6].

2-2 BUTTERWORTH FILTERS

Perhaps the simplest low-pass amplitude function is that of the *Butterworth filter* [33], defined in the nth-order case by

$$|H(j\omega| = \frac{A}{\sqrt{1 + (\omega/\omega_c)^{2n}}} \quad (n = 1, 2, 3, \ldots) \quad (2\text{-}1)$$

BUTTERWORTH FILTERS

The Butterworth response monotonically decreases (never increases) as the frequency increases. Also, the response improves as the order increases, as may be seen in Fig. 2-6, where several Butterworth responses are shown for the case $A = 1$.

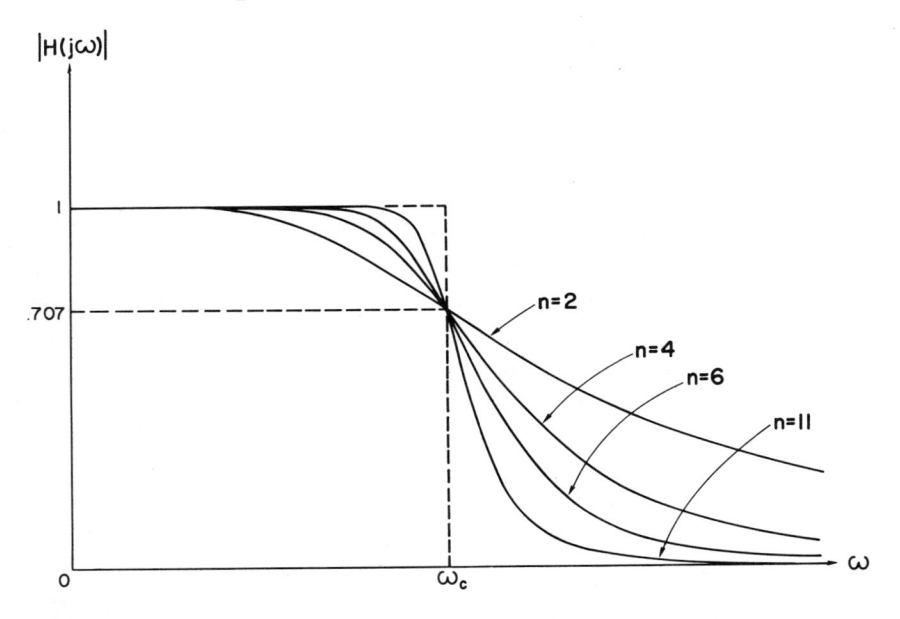

FIGURE 2-6 Low-pass Butterworth amplitude responses.

The Butterworth filter is an all-pole filter, having in the general case a transfer function of the type

$$\frac{V_2}{V_1} = \frac{Kb_0}{s^n + b_{n-1}s^{n-1} + \cdots + b_1 s + b_0} \quad (2\text{-}2)$$

where K is a constant. In the *normalized* case, $\omega_c = 1$ rad/s, the transfer function may be factored, for $n = 2, 4, 6, \ldots$, in the form

$$\frac{V_2}{V_1} = \prod_{k=1}^{n/2} \frac{A_k}{s^2 + a_k s + b_k} \quad (2\text{-}3)$$

or for $n = 3, 5, 7, \ldots$, in the form

$$\frac{V_2}{V_1} = \frac{A_0}{s + b_0} \prod_{k=1}^{(n-1)/2} \frac{A_k}{s^2 + a_k s + b_k} \quad (2\text{-}4)$$

In both cases the coefficients are given by $b_0 = 1$ and for $k = 1, 2, \ldots$, by

$$a_k = 2 \sin \frac{(2k-1)\pi}{2n}, \qquad b_k = 1 \qquad (2\text{-}5)$$

The gain of the Butterworth filter described by Eq. (2-2) is evidently K (the value of the transfer function at $s = 0$). If the filter is constructed by cascading sections corresponding to the factors of Eq. (2-3) or (2-4), then A_k and/or A_0 will be the section gain. The gain of the filter is then the product of the section gains.

The Butterworth response is the *flattest* in the vicinity of $\omega = 0$ of any nth-order all-pole filter and is called *maximally flat* for this reason. Thus for low frequencies the Butterworth response best approximates the ideal response. However, for frequencies near the cutoff point and in the stopband, the Butterworth filter is distinctly inferior to the Chebyshev filter, which will be discussed in the next section.

On the other hand, the phase response of the Butterworth filter is better (more nearly linear) than those of the Chebyshev, inverse Chebyshev, and elliptic filters of comparable order [32, 33]. This is consistent with the general rule for filters of this type—the better the amplitude, the poorer the phase, and vice versa.

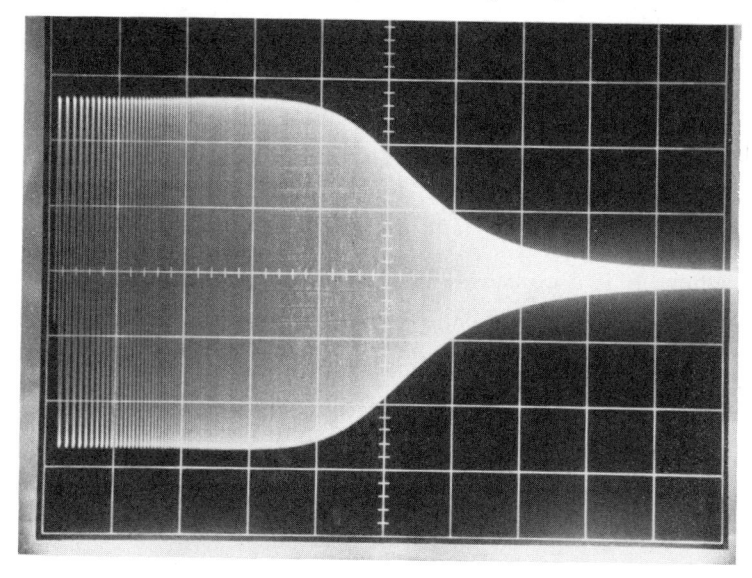

FIGURE 2-7 Response of an actual sixth-order Butterworth filter.

The normalized Butterworth filter transfer function (2-2) is given in Appendix A in the factored forms (2-3) and (2-4) for $n = 2, 3, \ldots, 10$.

The amplitude response of a sixth-order low-pass Butterworth filter that was constructed in the laboratory is shown in Fig. 2-7.

2-3 CHEBYSHEV FILTERS

The low-pass *Chebyshev* filter is the optimum all-pole filter, as noted earlier. It has amplitude response [32] given by

$$|H(j\omega)| = \frac{K}{\sqrt{1 + \epsilon^2 C_n^2(\omega/\omega_c)}} \qquad (n = 1, 2, 3, \ldots) \qquad (2\text{-}6)$$

The quantities ϵ and K are constants and C_n is the Chebyshev polynomial of the first kind of degree n, given by

$$C_n(x) = \cos(n \arccos x) \qquad (2\text{-}7)$$

The amplitude reaches its peak value of K at the points where C_n is zero. Since these points are distributed across the passband, the Chebyshev response has ripples in the passband and is monotonic elsewhere. The width of the ripples is determined by the value of ϵ and their number is determined by n. The value of K determines the gain of the filter. Several Chebyshev responses are shown in Fig. 2-8 for $K = 1$ and $\omega_c = 1$ rad/s.

The Chebyshev filter is sometimes called an *equiripple filter* because the ripples are all equal in magnitude. For $K = 1$, the case considered in Fig. 2-8, the ripple width is given by

$$\text{RW} = 1 - \frac{1}{\sqrt{1 + \epsilon^2}} \qquad (2\text{-}8)$$

Thus we may make RW as small as we like by making ϵ sufficiently small.

This constant ripple width is often expressed in dB by calculating the minimum allowable passband loss. This is given by

$$\begin{aligned}\alpha &= -20 \log_{10}\left(\frac{1}{\sqrt{1 + \epsilon^2}}\right) \\ &= 10 \log_{10}(1 + \epsilon^2)\end{aligned} \qquad (2\text{-}9)$$

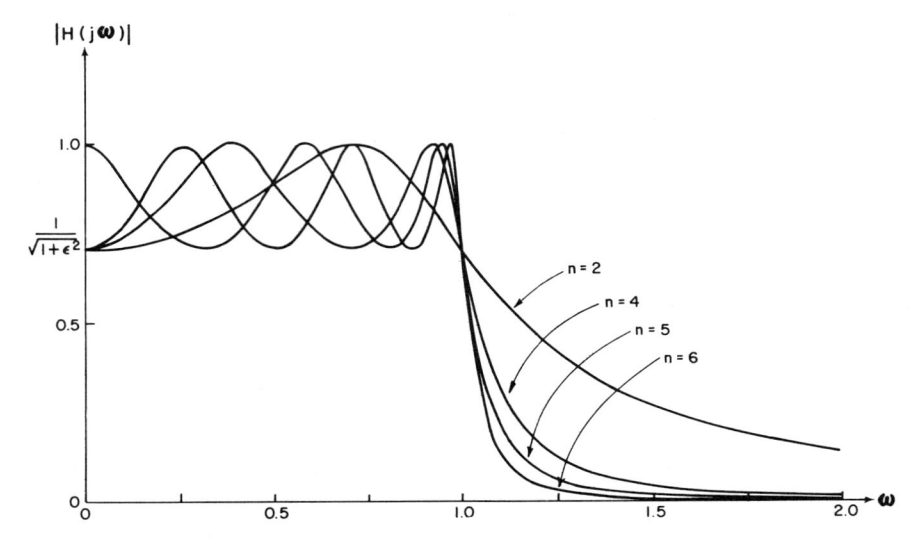

FIGURE 2-8 Low-pass Chebyshev amplitude responses.

and may be used to characterize the Chebyshev filter. For example, a $\frac{1}{2}$-dB filter is one with ϵ such that $\alpha = \frac{1}{2}$ (requiring that $\epsilon = 0.3493$). In the general case, Eq. (2-9) may be solved for ϵ to give

$$\epsilon = \sqrt{10^{\alpha/10} - 1} \qquad (2\text{-}10)$$

The widest allowable ripple width is that of the 3-dB Chebyshev filter for which $\epsilon = 1$ in Eq. (2-9). (More exactly, since log 2 is not precisely 0.3, we need $\epsilon = 0.99763$.)

Comparing Figs. 2-1 and 2-8, we have $A = 1$ and $A_1 = 1/\sqrt{1+\epsilon^2}$. For a given case, A_2 may also be specified, which would determine the value of ω_1. The frequency $\omega_c = 1$ rad/s is the cutoff point or the terminal point of the ripple channel. If we are interested instead in ω_{3dB}, the point at which the response is down 3 dB, we have [16]

$$\omega_{3\,dB} = \cosh\left(\frac{1}{n} \operatorname{arccosh} \frac{1}{\epsilon}\right) \qquad (2\text{-}11)$$

We note that $\omega_c = \omega_{3\,dB}$ if $\epsilon = 1$, in which case we have the 3-dB Chebyshev filter.

The transfer functions of low-pass Chebyshev filters are identical in form to those of the Butterworth filter given previously

SEC. 2-3   CHEBYSHEV FILTERS   19

in Eqs. (2-2), (2-3), and (2-4). The denominator polynomials in the case of the factored forms (2-3) and (2-4) are tabulated for $\omega_c = 1$ rad/s in Appendix A for $n = 2, 3, \ldots, 10$. The passband ripple cases given are 0.1, 0.5, 1, 2, and 3 dB.

The amplitude response of the Chebyshev filter for a given order is better than that of the Butterworth in the sense that the Chebyshev transition width is smaller. The Chebyshev phase response, however, is inferior (more nonlinear) to that of the Butterworth. Chebyshev phase responses for $n = 2$ through 7 are shown in Fig. 2-9. For comparison purposes, the phase response of a sixth-order Butterworth filter is shown dashed in Fig. 2-9. We may note also that phase responses of higher-order Chebyshev filters are inferior to those of lower order. This is

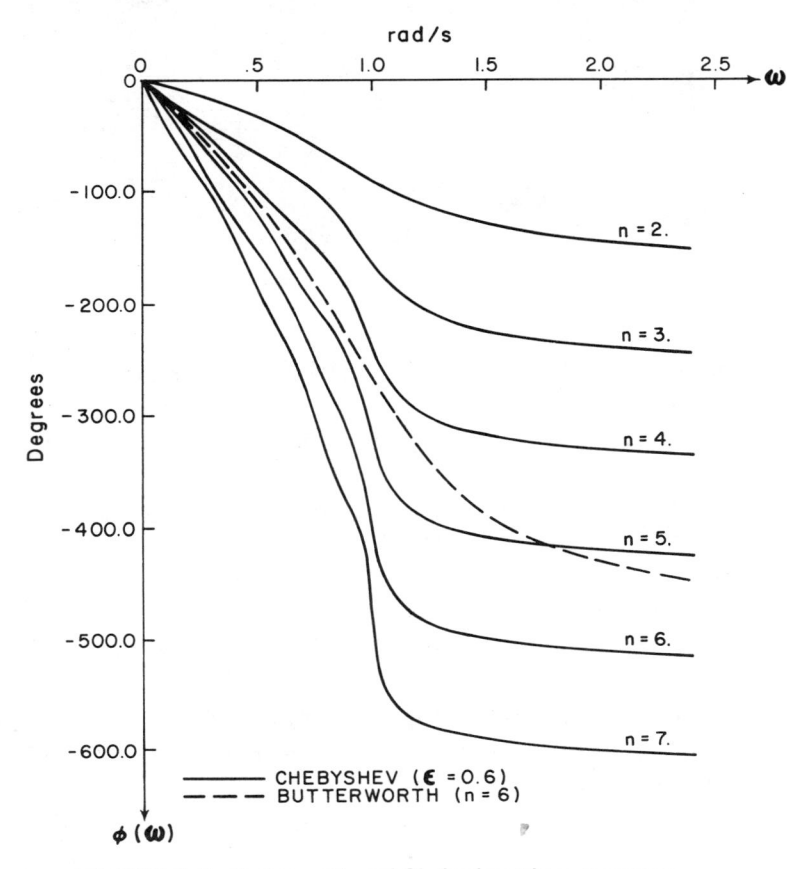

FIGURE 2-9 Butterworth and Chebyshev phase responses.

consistent with the fact that higher-order Chebyshev amplitudes are superior to those of lower order.

The amplitude response of an actual fourth-order 1-dB Chebyshev filter is shown in Fig. 2-10.

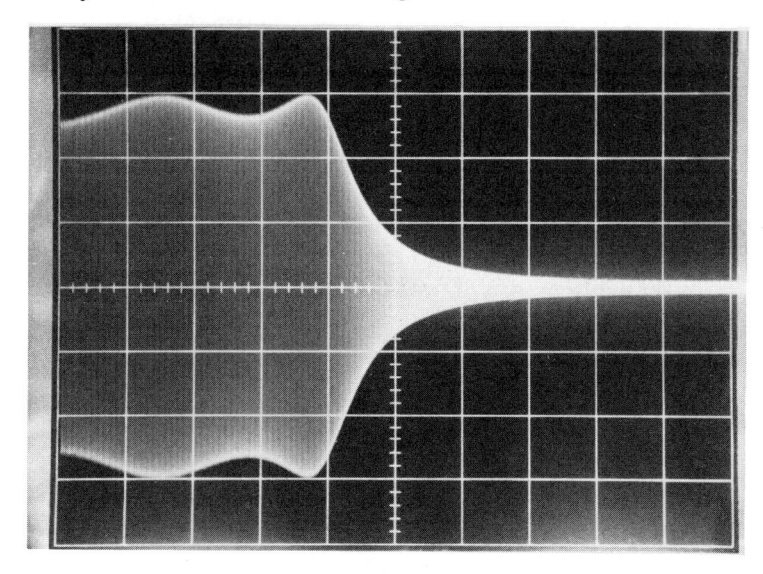

FIGURE 2-10 *Response of an actual fourth-order Chebyshev filter.*

2-4 SELECTION OF THE MINIMUM ORDER

As we may see from Figs. 2-6 and 2-8, the higher the order of the Butterworth and Chebyshev filters, the better the response. However, higher order is accompanied by more complex circuitry and, therefore, higher cost. Thus it is of interest to the designer to determine the minimum order required of the filter to satisfy the given constraints.

In other words, suppose that in the general case of Fig. 2-2 we are given the maximum allowable passband loss α_1 (in dB), the minimum allowable stopband loss α_2 (dB), the cutoff frequency ω_c (rad/s) or f_c (Hz), and the maximum allowable transition width TW, given by

$$\text{TW} = \omega_1 - \omega_c \qquad (2\text{-}12)$$

(That is, the stopband is to start with some $\omega_2 \leq \omega_1$.) The problem is to find the minimum order n that will satisfy all these conditions.

SEC. 2-4　SELECTION OF THE MINIMUM ORDER

In the case of the Butterworth filter with $\alpha_1 = 3$ dB, the minimum order may be found by applying the preceding conditions to Eq. (2-1) and solving for n. The result is given by

$$n = \frac{\log(10^{\alpha_2/10} - 1)}{2 \log(\omega_1/\omega_c)} \qquad (2\text{-}13)$$

where the logs may be either natural logs or base 10. From Eq. (2-12) we may write

$$\frac{\omega_1}{\omega_c} = \frac{TW}{\omega_c} + 1 \qquad (2\text{-}14)$$

which may be used in Eq. (2-13) to relate n to the transition width rather than to ω_1. The quantity TW/ω_c is the *normalized* transition width and is dimensionless. Therefore, TW and ω_c may be given in rad/s or in Hz.

In a similar manner, using Eq. (2-6) for $K = 1$, the minimum order for a Chebyshev filter is given by

$$n = \frac{\operatorname{arccosh}\sqrt{(10^{\alpha_2/10} - 1)/(10^{\alpha_1/10} - 1)}}{\operatorname{arccosh}(\omega_1/\omega_c)} \qquad (2\text{-}15)$$

Again, Eq. (2-14) may be used to eliminate ω_1.

As an example, suppose that we are given $\alpha_1 = 3$ dB, $\alpha_2 = 20$ dB, $f_c = 1000$ Hz, and the transition band is not to exceed $TW = 300$ Hz. From Eq. (2-14) we have

$$\frac{\omega_1}{\omega_c} = \frac{300}{1000} + 1 = 1.3$$

so that by Eq. (2-13), a Butterworth filter satisfying these requirements must have a minimum order given by

$$n = \frac{\log(10^2 - 1)}{2 \log 1.3}$$
$$= 8.76$$

Since the order must be an integer, we must use $n = 9$.

A Chebyshev filter satisfying these requirements has a minimum order given by Eq. (2-15) as

$$n = \frac{\operatorname{arccosh}\sqrt{(10^2 - 1)/(2 - 1)}}{\operatorname{arccosh} 1.3}$$
$$= 3.95$$

Again, using the next higher integral value, we have $n = 4$.

This example dramatically illustrates the superiority of the Chebyshev filter over the Butterworth filter if the prime consideration is the amplitude response. The Chebyshev filter in this case performs the same function as a Butterworth filter of more than twice the complexity.

Equations (2-13) and (2-15) may also be used to find the transition width TW for Butterworth and Chebyshev filters of a fixed order. For example, substituting Eq. (2-14) in Eq. (2-13) and solving for TW/ω_c results in

$$\frac{TW}{\omega_c} = \sqrt[2n]{10^{\alpha_2/10} - 1} - 1 \qquad (2\text{-}16)$$

for the Butterworth filter. Repeating the procedure with Eq. (2-15) we have, for the Chebyshev filter,

$$\frac{TW}{\omega_c} = \cosh\left(\frac{1}{n} \operatorname{arccosh}\sqrt{\frac{10^{\alpha_2/10} - 1}{10^{\alpha_1/10} - 1}}\right) - 1 \qquad (2\text{-}17)$$

To illustrate the use of these formulas, let us find the transition width TW of the Butterworth filter of the previous example having $\alpha_1 = 3$ dB (the only Butterworth case we are considering), $\alpha_2 = 20$ dB, $f_c = 1000$ Hz, and $n = 9$. From Eq. (2-16) we have

$$\frac{TW}{\omega_c} = \sqrt[18]{99} - 1 = 0.291$$

and thus TW in rad/s is $0.291\omega_c$ or in Hz is $0.291 f_c = 291$. This is consistent with the previous example, in which it was shown that $TW \leq 300$ Hz.

2-5 INFINITE-GAIN MULTIPLE-FEEDBACK LOW-PASS FILTERS

There are many ways of constructing Butterworth and Chebyshev low-pass active filters. In the remaining sections of this chapter we shall consider some of the more common circuits in current use, beginning with the simplest (from the standpoint of the number of circuit elements required) and proceeding to the most sophisticated.

SEC. 2-5 MULTIPLE-FEEDBACK LOW-PASS FILTERS

In the case of second-order low-pass filters with cutoff frequency ω_c rad/s, the typical all-pole transfer function is given by

$$\frac{V_2}{V_1} = \frac{KC\omega_c^2}{s^2 + B\omega_c s + C\omega_c^2} \qquad (2\text{-}18)$$

The constants B and C are the normalized coefficients, since for $\omega_c = 1$ the transfer function reduces to the form of Eq. (2-2) with $n = 2$. These coefficients are given in Appendix A in the case of Butterworth or Chebyshev filters. The constant K is the gain, which, of course, must be specified.

In the case of higher-order filters, Eq. (2-18) is the transfer function of a typical second-order stage and K is the stage gain. The coefficients B and C are the stage coefficients given in Appendix A.

One of the simplest active filter networks that realizes the low-pass transfer function Eq. (2-18) is that of Fig. 2-11 [3]. This

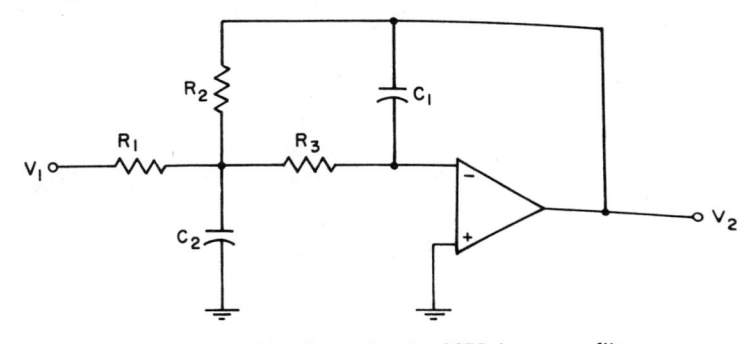

FIGURE 2-11 *Second-order MFB low-pass filter.*

circuit is sometimes called an *infinite-gain multiple-feedback* (MFB) *circuit* because of the two feedback paths through C_1 and R_2, and because the op amp is serving as an infinite-gain device rather than a finite-gain device. (An example of the latter type will be considered in the next section.) The circuit realizes Eq. (2-18) with an *inverting* gain $-K$ ($K > 0$) and

$$C\omega_c^2 = \frac{1}{R_2 R_3 C_1 C_2}$$

$$B\omega_c = \frac{1}{C_2}\left(\frac{1}{R_1} + \frac{1}{R_2} + \frac{1}{R_3}\right) \qquad (2\text{-}19)$$

$$K = \frac{R_2}{R_1}$$

Resistance values satisfying Eq. (2-19) are given by

$$R_2 = \frac{2(K+1)}{[BC_2 + \sqrt{B^2 C_2^2 - 4CC_1 C_2 (K+1)}]\omega_c}$$

$$R_1 = \frac{R_2}{K} \qquad (2\text{-}20)$$

$$R_3 = \frac{1}{CC_1 C_2 \omega_c^2 R_2}$$

where C_1 and C_2 are arbitrary. The resistances are in ohms and the capacitances are in farads.

Therefore, for a given K, B, C, and ω_c, we may select values of C_1 and C_2 and calculate the required resistance values. The capacitances should be standard values that result in a real value of R_2. This is the case if we have

$$C_1 \leq \frac{B^2 C_2}{4C(K+1)} \qquad (2\text{-}21)$$

A good rule of thumb is to select a standard value of C_2 near $10/f_c$ μF and select the largest available standard value of C_1 that satisfies Eq. (2-21). The resistances should be close to the calculated values in Eq. (2-20). The higher the order of the filter, the more critical this requirement is. If the calculated resistances are not available, it is worth noting that they may all be multiplied by a common factor provided that the capacitances are divided by the same factor.

As an example, suppose that we wish to construct a second-order MFB 0.5-dB Chebyshev filter with a passband of 1000 Hz and a gain of 2. In this case we have $K = 2$, $\omega_c = 2\pi(1000)$, and from Appendix A, $B = 1.425625$ and $C = 1.516203$. Choosing $C_2 = 10/f_c = 10/1000 = 0.01$ μF $= 10^{-8}$ F, we have, by Eq. (2-21),

$$C_1 \leq \frac{(1.425625)^2 (0.01)}{4(1.516203)(3)} = 0.0011 \ \mu\text{F}$$

We select $C_1 = 0.001$ μF $= 1$ nF, a standard value, and calculate the resistances by Eq. (2-20). The results are:

$$R_2 = \frac{2(3)}{[(1.425625)(10^{-8}) + \sqrt{(1.425625)^2 (10^{-16}) - 4(1.516203)(10^{-9})(10^{-8})(3)}](2000\pi)}$$

SEC. 2-5 MULTIPLE-FEEDBACK LOW-PASS FILTERS

$$= 0.506 \times 10^5 \, \Omega$$
$$= 50.6 \, \text{k}\Omega$$
$$R_1 = \frac{50.6}{2} = 25.3 \, \text{k}\Omega$$

and

$$R_3 = \frac{1}{(1.516203)(10^{-9})(10^{-8})(2000\pi)^2(50.6)(10^3)}$$
$$= 0.330 \times 10^5 \, \Omega$$
$$= 33 \, \text{k}\Omega$$

As a final example, suppose that we want a sixth-order MFB Butterworth filter with $f_c = 1000$ Hz and a gain of 8. There will be three second-order stages, each with a transfer function like Eq. (2-18). Let us choose each section to have a gain $K = 2$, which yields the required filter gain of $2 \times 2 \times 2 = 8$. By Appendix A we have for the first stage, $B = 0.517638$ and $C = 1$. Again, we shall select $C_2 = 0.01 \, \mu\text{F}$, in which case Eq. (2-21) becomes $C_1 \leq 0.00022 \, \mu\text{F}$. We choose $C_1 = 200$ pF and obtain from Eq. (2-20) the resistances, given by

$$R_2 = 139.4 \, \text{k}\Omega$$
$$R_1 = 69.7 \, \text{k}\Omega$$
$$R_3 = 90.9 \, \text{k}\Omega$$

The other two stages are obtained in the same manner and the three stages cascaded to obtain the sixth-order Butterworth filter. The resulting circuit has the amplitude response shown earlier in Fig. 2-7.

Because of its relative simplicity, the MFB filter is one of the more popular inverting-gain types. It also has the advantages of good stability characteristics and low output impedance [9]; thus it can be readily cascaded with other sections to form a higher-order filter. A disadvantage of the circuit is that it cannot attain a high pole-pair quality factor Q without large spreads of element values and high sensitivities to changes in the element values. For best results the gain K and the quality factor Q should both be limited to approximately 10. The gain could be higher if Q is lower, maintaining a limit of, say, $KQ = 100$, with $Q \leq 10$.

As we may recall from Eq. (1-11), the pole-pair quality factor is given by $Q = \sqrt{C}/B$. The highest Q stage in the sixth-order low-pass Butterworth filter, by Appendix A, is $Q = 1/0.517638 = 1.93$ of the first stage. Thus the MFB filter may be used with reasonably good results in this example.

The MFB filter design procedure is summarized and practical suggestions are given in Section 2-10.

2-6 VCVS LOW-PASS FILTERS

A widely used second-order low-pass filter circuit which yields a noninverting (positive) gain is that of Fig. 2-12 [26]. This circuit is

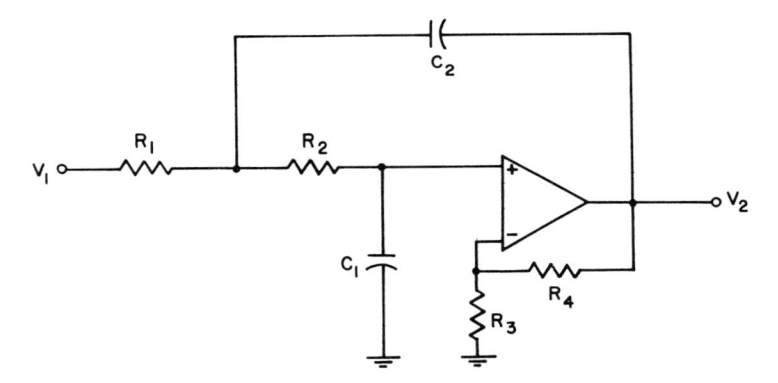

FIGURE 2-12 Second-order VCVS low-pass filter.

sometimes called a *VCVS filter* because the op amp and its two connected resistors R_3 and R_4 form a *voltage-controlled voltage source* (VCVS).

This circuit realizes the second-order low-pass function (2-18) with

$$C\omega_c^2 = \frac{1}{R_1 R_2 C_1 C_2}$$

$$B\omega_c = \frac{1}{C_2}\left(\frac{1}{R_1} + \frac{1}{R_2}\right) + \frac{1}{R_2 C_1}(1 - \mu) \qquad (2\text{-}22)$$

$$K = \mu = 1 + \frac{R_4}{R_3}$$

SEC. 2-6 VCVS LOW-PASS FILTERS

The quantity $\mu \geq 1$ is the gain of the VCVS, as well as the filter gain. Resistance values satisfying Eq. (2-22) are given by

$$R_1 = \frac{2}{[BC_2 + \sqrt{[B^2 + 4C(K-1)]C_2^2 - 4CC_1C_2}]\omega_c}$$

$$R_2 = \frac{1}{CC_1C_2R_1\omega_c^2}$$

$$R_3 = \frac{K(R_1 + R_2)}{K - 1} \quad (K \neq 1)$$

$$R_4 = K(R_1 + R_2)$$

(2-23)

where C_1 and C_2 are arbitrary. The resistances R_3 and R_4 are selected in such a way as to minimize the dc *offset* of the op amp. (Recall that ideally the *offset* voltage between the input terminals should be zero.)

If we desire $K = 1$, then R_1 and R_2 are given by Eq. (2-23), but in that case we have $R_3 = \infty$ (open circuit) and $R_4 = 0$ (short circuit). For minimum dc offset, $R_4 = R_1 + R_2$, but in most noncritical operations, a short circuit will suffice. In this case the VCVS is a voltage *follower*. The output voltage is equal to, or *follows*, the input voltage.

The design of the VCVS filter is performed as in the case of the MFB filter of Section 2-5. We select standard values of C_2, preferably near $10/f_c$ μF, and of C_1, satisfying the inequality

$$C_1 \leq \frac{[B^2 + 4C(K-1)]C_2}{4C} \quad (2\text{-}24)$$

(This assures us that R_1 is real.) The resistances are then found from Eq. (2-23), with the modification given above if $K = 1$.

The VCVS circuit achieves a noninverting gain, as pointed out earlier, with a minimal number of elements. (It requires only one more resistor than the MFB filter.) It has a low output impedance, a low spread of element values, and a capability of relatively high gains. In addition, it is relatively easy to tune. The gain may be set precisely, for example, by adjusting R_3 and R_4 by means of a potentiometer. Like the MFB filter, however, the VCVS filter should be used for pole-pair Q's of 10 or less.

A complete design procedure is given in Section 2-11.

2-7 BIQUAD LOW-PASS FILTERS

The last second-order low-pass filter we consider, which realizes the transfer function (2-18), is the *biquad* circuit of Fig. 2-13 [31].

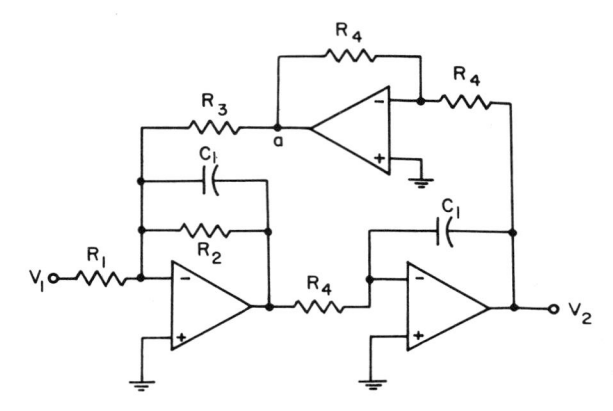

FIGURE 2-13 *Second-order biquad low-pass filter.*

This circuit has more elements than the MFB and VCVS circuits of previous sections, but it is a far superior circuit because of its tuning advantages and excellent stability. Pole-pair Q's of up to 100 may be readily attained, and it is relatively easy to cascade several biquad sections to obtain higher-order filters.

The circuit realizes Eq. (2-18) with noninverting gain K and

$$C\omega_c^2 = \frac{1}{R_3 R_4 C_1^2}$$

$$B\omega_c = \frac{1}{R_2 C_1} \quad (2\text{-}25)$$

$$K = \frac{R_3}{R_1}$$

The resistance values are given by

$$R_1 = \frac{1}{KCC_1^2\omega_c^2 R_4}$$

$$R_2 = \frac{1}{BC_1\omega_c} \quad (2\text{-}26)$$

$$R_3 = \frac{1}{CC_1^2\omega_c^2 R_4}$$

SEC. 2-8   TUNING THE SECOND-ORDER FILTERS   29

with C_1 and R_4 arbitrary. If C_1 is chosen near $10/f_c$ μF, a reasonable value of R_4 is

$$R_4 = \frac{1}{\omega_c C_1} \tag{2-27}$$

in which case we have

$$R_1 = \frac{R_4}{KC}$$
$$R_2 = \frac{R_4}{B} \tag{2-28}$$
$$R_3 = \frac{R_4}{C}$$

The biquad circuit is relatively easy to tune, as we may see from Eq. (2-28). For a given R_4, varying R_2 affects B and varying R_3 affects C. Then with C correctly fixed, varying R_1 affects K. If an inverting gain is desired, the output V_2 may be taken at node a, keeping the element values the same as before.

A complete design procedure is given in Section 2-12.

2-8 TUNING THE SECOND-ORDER FILTERS

The tuning of a second-order filter or a second-order stage of a higher-order filter can be much more easily accomplished if the designer knows what the general shape of the response should be. In the case of the second-order low-pass function (2-18), the amplitude will have a peak of K_m occurring at a frequency f_m Hz, provided that $B^2/C < 2$. This response is shown in Fig. 2-14(a) and the values of K_m and f_m are given by

$$K_m = \frac{2CK}{B\sqrt{4C - B^2}} \tag{2-29}$$

$$f_m = f_c \sqrt{C - \frac{B^2}{2}} \tag{2-30}$$

In terms of the pole-pair Q (given by $Q = \sqrt{C}/B$), peaking occurs, as in Fig. 2-14(a), if $Q > 1/\sqrt{2} = 0.707$. If $Q \leq 0.707$ (or $B^2/C \geq 2$), there is no peaking, and the response is as shown in Fig. 2-14(b). In both figures, f_c is the cutoff frequency of the

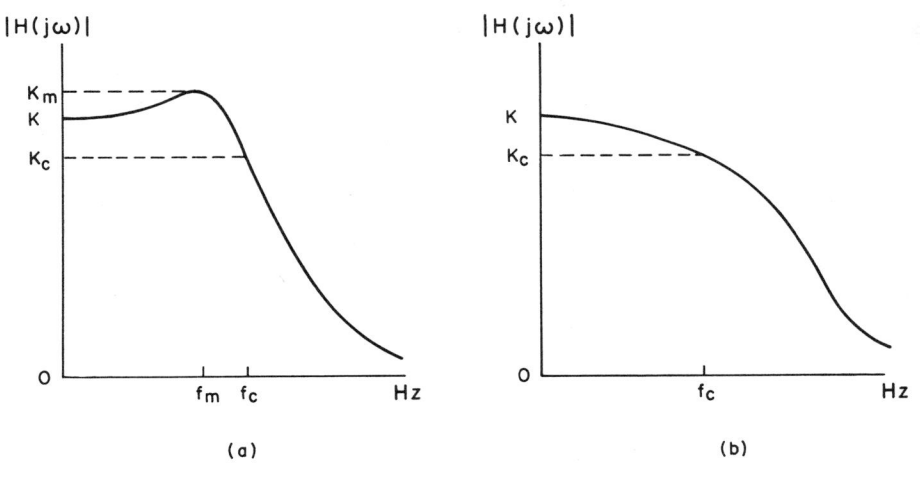

FIGURE 2-14 Low-pass amplitude responses for: (a) $Q > 0.707$ and (b) $Q \leq 0.707$.

filter, and its corresponding amplitude is given by

$$K_c = \frac{KC}{\sqrt{(C-1)^2 + B^2}} \qquad (2\text{-}31)$$

As an example, let us consider the fourth-order Butterworth filter with $f_c = 1000$ Hz and a gain per section of $K = 2$. By Appendix A we have, for the first stage, $B = 0.765367$ and $C = 1$. Thus from Eq. (2-29) we have

$$K_m = \frac{2(1)(2)}{0.765367\sqrt{4(1) - (0.765367)^2}}$$
$$= 2.8284$$

and from Eq. (2-30) we have

$$f_m = 1000\sqrt{1 - \frac{(0.765367)^2}{2}}$$
$$= 841 \text{ Hz}$$

At f_c, which in this case is the 3-dB point, we have, by Eq. (2-31), the amplitude,

$$K_c = \frac{2(1)}{\sqrt{(1-1)^2 + (0.765367)^2}}$$
$$= 2.6131$$

Therefore, the amplitude should resemble Fig. 2-14(a) (since $Q = 1/0.765367 = 1.31$), with a peak amplitude of 2.8284 occurring at 841 Hz and an amplitude of 2.6131 occurring at 1000 Hz. The dc value of the amplitude is, of course, $K = 2$.

For the second stage we have $B = 1.847759$ and $C = 1$. Therefore, $Q = 1/B = 0.54$ and the response will look like Fig. 2-14(b) with $K = 2$ and a 1000-Hz amplitude of

$$K_c = \frac{2(1)}{\sqrt{(1-1)^2 + (1.847759)^2}} = 1.0824$$

As a check, when the two stages are cascaded, the amplitude at 0 Hz is $2 \times 2 = 4$ and that at 1000 Hz is $2.6131 \times 1.0824 = 2.828$. The latter value is 0.707×4, as it should be.

Tuning suggestions are given in the design procedures for the various filter types in Sections 2-10, 2-11, and 2-12.

2-9 ODD-ORDER FILTERS

In the case of odd-order Butterworth or Chebyshev filters, one stage must have a first-order transfer function like the first factor in Eq. (2-4). In the case of a general cutoff frequency $\omega_c = 2\pi f_c$ rad/s, this first-order factor is of the form

$$\frac{V_2}{V_1} = \frac{KC\omega_c}{s + C\omega_c} \tag{2-32}$$

where K is the stage gain and C is given as the stage 1 coefficient in Appendix A.

A circuit that achieves Eq. (2-32) for a gain $K > 1$ is shown in Fig. 2-15. The capacitance C_1 is to be chosen, preferably near $10/f_c$ μF, and the resistances are given by

$$R_1 = \frac{1}{\omega_c C_1 C}$$
$$R_2 = \frac{KR_1}{K-1} \tag{2-33}$$
$$R_3 = KR_1$$

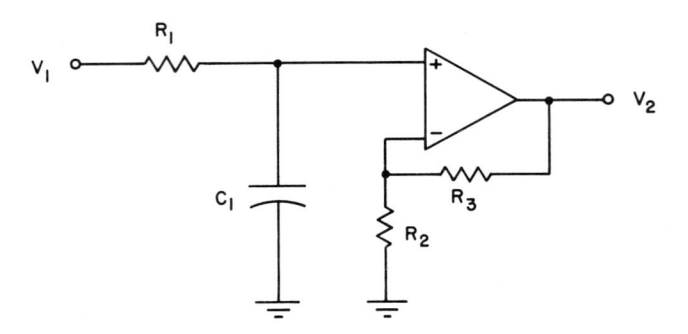

FIGURE 2-15 First-order low-pass filter circuit.

If a gain of $K = 1$ is desired, we may use Fig. 2-16 as the first-order stage. In this case R_1 is as given in Eq. (2-33) and C_1 is again arbitrary.

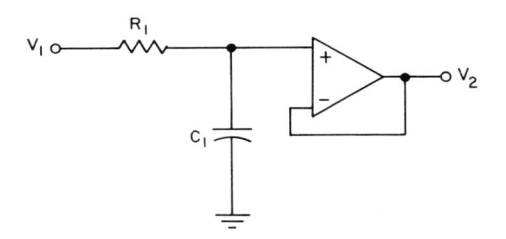

FIGURE 2-16 First-order low-pass section with unity gain.

As an example, suppose that we want a third-order Butterworth filter with $f_c = 1000$ Hz and a gain of 2. By Appendix A the first-order stage has $C = 1$ in Eq. (2-32), and the second-order stage has $B = C = 1$ in Eq. (2-18). We shall take the first-order stage gain as $K = 1$ and the second-order stage gain as $K = 2$. Thus the first-order stage is the circuit of Fig. 2-16. Choosing $C_1 = 0.01$ μF, we have, by the first of Eq. (2-33),

$$R_1 = \frac{1}{(2000\pi)(10^{-8})(1)} \Omega = 15.915 \text{ k}\Omega$$

We may realize the second-order stage by the procedure of Section 2-6 or 2-7 for a noninverting gain, or Section 2-5 for an inverting gain.

2-10 MFB LOW-PASS FILTER DESIGN SUMMARY

To design a second-order low-pass filter, or a second-order stage of a higher-order filter having a given cutoff frequency f_c Hz (or $\omega_c = 2\pi f_c$ rad/s), gain K, and of Butterworth or Chebyshev type, perform the following steps.

1. Find the normalized coefficients B and C from the appropriate table of Appendix A

2. Select a standard value of C_2 (preferably near $10/f_c$ μF) and a standard value of C_1 satisfying

$$C_1 \leq \frac{B^2 C_2}{4C(K+1)}$$

(preferably the largest available standard value). Calculate the resistance values given by

$$R_2 = \frac{2(K+1)}{[BC_2 + \sqrt{B^2 C_2^2 - 4CC_1 C_2(K+1)}]\omega_c}$$

$$R_1 = \frac{R_2}{K}$$

$$R_3 = \frac{1}{CC_1 C_2 \omega_c^2 R_2}$$

3. Select standard values of resistance as close as possible to the calculated values and construct the filter, or its second-order stages, in accordance with Fig. 2-17

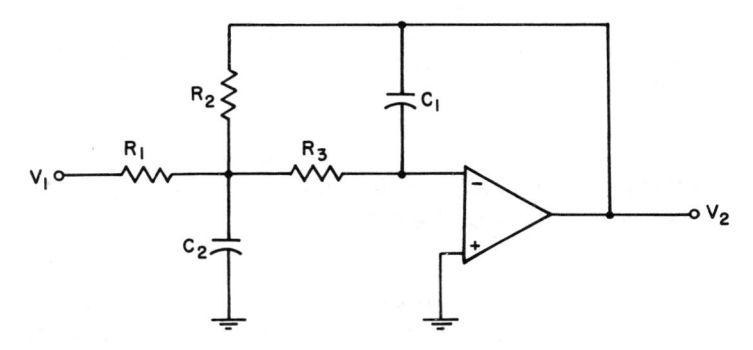

FIGURE 2-17 MFB low-pass filter circuit.

Comments

(a) For best performance, element values close to those selected and calculated should be used. Higher-order filters require more accurate element values than lower-order filters. The performance of the filter is unchanged if all the resistances are multiplied and the capacitances divided by a common factor

(b) The input impedance of the op amp should be at least $10 R_{eq}$, where

$$R_{eq} = R_3 + \frac{R_1 R_2}{R_1 + R_2}$$

The open-loop gain of the op amp should be at least 50 times the amplitude of the filter, or stage, at f_c, and its slew rate (volts per microsecond) should be at least $\frac{1}{2}\omega_c \times 10^{-6}$ times the peak-to-peak output voltage

(c) The gain of each stage is an inverting gain with magnitude $K = R_2/R_1$. Thus the gain may be set by using a potentiometer in lieu of resistor R_2. Resistor R_3 may be varied to adjust f_c, after which, in the case of amplitude peaking, R_1 may be varied to adjust f_m (see Section 2-8). These steps may be repeated if necessary

(d) The circuit should be used only for filters or filter stages with gain K and stage pole-pair Q ($\sqrt{C/B}$) of 10 or less. The gain could be higher if Q is lower, with limits of $KQ = 100$ and $Q = 10$

(e) The order required for a given transition width, or conversely the transition width ensuing from a given order, may be found as described in Section 2-4

An example of an MFB low-pass filter design was given in Section 2-5.

2-11 VCVS LOW-PASS FILTER DESIGN SUMMARY

To design a second-order low-pass filter, or a second-order stage of a higher-order filter, having a given cutoff frequency f_c Hz (or $\omega_c = 2\pi f_c$ rad/s), gain K, and of Butterworth or Chebyshev type, perform the following steps.

SEC. 2-11   VCVS LOW-PASS FILTER DESIGN

1. Find the normalized coefficients B and C from the appropriate table of Appendix A
2. Select a standard value of C_2 (preferably near $10/f_c$ μF) and a standard value of C_1 satisfying

$$C_1 \leq \frac{[B^2 + 4C(K-1)]C_2}{4C}$$

(preferably the largest available standard value). If $K > 1$, calculate the resistance values given by

$$R_1 = \frac{2}{[BC_2 + \sqrt{[B^2 + 4C(K-1)]C_2^2 - 4CC_1C_2}]\omega_c}$$

$$R_2 = \frac{1}{CC_1C_2R_1\omega_c^2}$$

$$R_3 = \frac{K(R_1 + R_2)}{K-1}$$

$$R_4 = K(R_1 + R_2)$$

If $K = 1$, then R_1 and R_2 are as above, but R_3 is replaced by an open circuit and R_4 by a short circuit

3. Select standard values of resistance as close as possible to the calculated values and construct the filter, or its stages, in accordance with Fig. 2-18

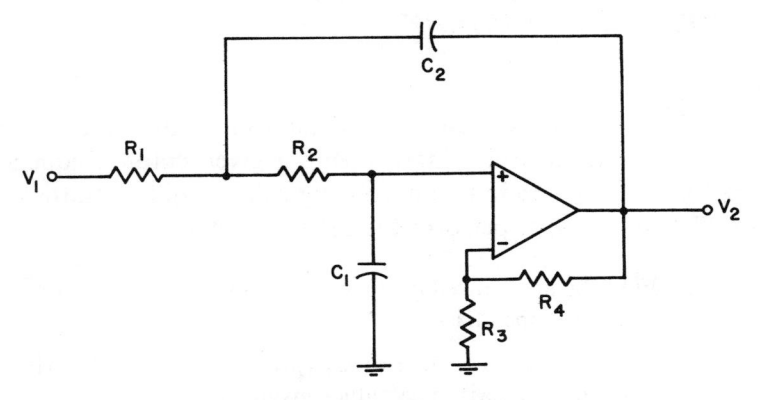

FIGURE 2-18 VCVS low-pass filter circuit.

Comments

(a) Comments (a), (b), (d), and (e) for the MFB filter of Section 2-10 apply directly, except that in (b), $R_{eq} = R_1 + R_2$

(b) The values of R_3 and R_4 are chosen to minimize the dc offset of the op amp. The section gain is a noninverting one given by

$$K = 1 + \frac{R_4}{R_3}$$

so other values of R_3 and R_4 may be used as long as their ratio is maintained

(c) There must be a dc return to ground at the filter input. Therefore, the stage must not be capacitively coupled to a source or another stage at node V_1

(d) The gain K may be set by using a potentiometer in lieu of resistors R_3 and R_4 with the center tap connected to the inverting input of the op amp. Resistors R_1 and R_2 may be changed by equal percentages to adjust f_c without affecting the pole-pair Q (see Section 2-8). These steps may be repeated if necessary

The VCVS low-pass filter was discussed in Section 2-6.

2-12 BIQUAD LOW-PASS FILTER DESIGN SUMMARY

To design a second-order low-pass filter, or a second-order stage of a higher-order filter, having a given cutoff frequency f_c Hz (or $\omega_c = 2\pi f_c$ rad/s), noninverting gain K, and of Butterworth or Chebyshev type, perform the following steps.

1. Find the normalized coefficients B and C from the appropriate table of Appendix A
2. Select a standard value of C_1 (preferably near $10/f_c$ μF), and calculate the resistance values given by

$$R_4 = \frac{1}{\omega_c C_1}$$

$$R_1 = \frac{R_4}{KC}$$

SEC. 2-12 BIQUAD LOW-PASS FILTER DESIGN 37

$$R_2 = \frac{R_4}{B}$$

$$R_3 = \frac{R_4}{C}$$

[Alternatively, R_4 may be arbitrary, with the other resistances given by Eq. (2-26).]

3. Select standard values of resistance as close as possible to the calculated values and construct the filter, or its stages, in accordance with Fig. 2-19

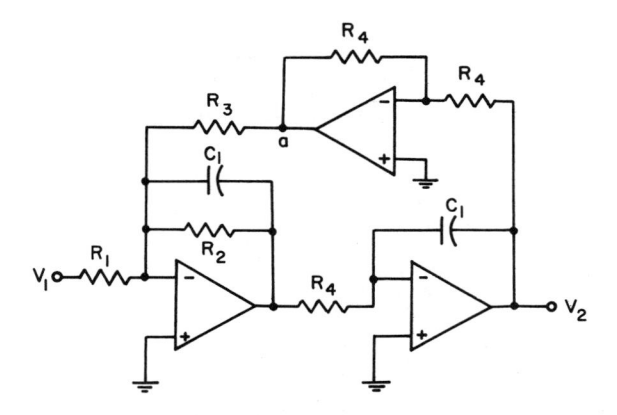

FIGURE 2-19 Biquad low-pass filter circuit.

Comments

(a) Comments (a), (b), and (e) for the MFB filter of Section 2-10 apply directly, except that in (b), R_{eq} for each op amp is the resistance R_1 or R_4 connected to its inverting input terminal

(b) The stage gain is $K = R_3/R_1$. If an inverting gain of $-K$ is desired, the output V_2 may be taken at point a

(c) Tuning may be accomplished by varying R_1 to adjust K, varying R_3 to adjust f_c, and varying R_2 to adjust the passband response (see Section 2-8). These steps may be repeated if necessary

(d) The filter may be used for stage pole-pair Q values up to 100

The biquad low-pass filter was discussed in Section 2-7.

2-13 ODD-ORDER LOW-PASS FILTER DESIGN SUMMARY

To design a first-order filter or a first-order stage of a higher odd-order filter having a given cutoff frequency f_c Hz (or $\omega_c = 2\pi f_c$ rad/s), gain K, and of Butterworth or Chebyshev type, perform the following steps.

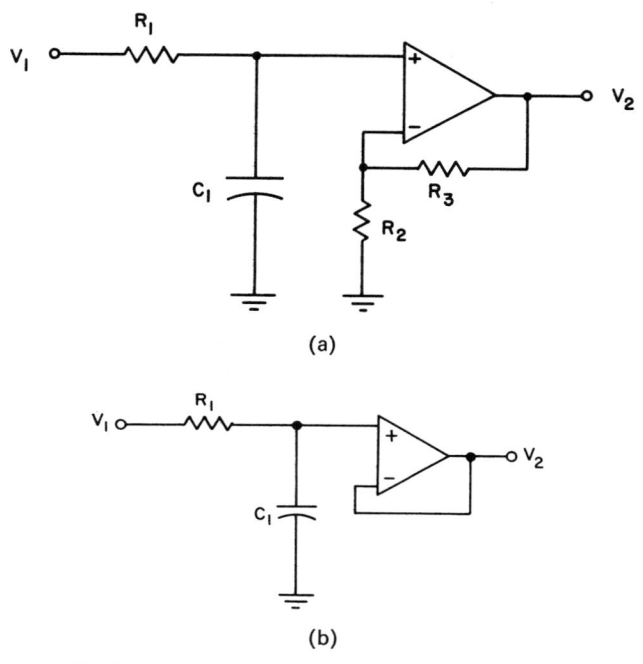

FIGURE 2-20 First-order low-pass filter circuits.

1. Find the normalized coefficient C for the first-order stage from the appropriate table of Appendix A
2. Select a standard value of C_1 (preferably near $10/f_c$ μF)
3. (a) If $K > 1$, use the circuit of Fig. 2-20(a) with resistance values given by

$$R_1 = \frac{1}{\omega_c C_1 C}$$

$$R_2 = \frac{KR_1}{K-1}$$

$$R_3 = KR_1$$

38

SEC. 2-13 ODD-ORDER LOW-PASS FILTER DESIGN

    (b) If $K = 1$, use the circuit of Fig. 2-20(b) with R_1 as given in 3(a)

4. The second-order stages of the odd-order filter may be constructed as indicated in Section 2-10, 2-11, or 2-12, and cascaded with the first-order section to form the filter

Comments

(a) Comments (a) and (b) for the MFB filter of Section 2-10 apply directly, except that in (b), $R_{eq} = R_1$

(b) The values of R_2 and R_3 in Fig. 2-20(a) are chosen to minimize the dc offset of the op amp. Other values may be used as long as their ratio is maintained, so

$$K = 1 + \frac{R_3}{R_2}$$

(c) There must be a dc return to ground at the filter input

(d) The gain K may be adjusted in Fig. 2-20(a) by using a potentiometer in lieu of resistors R_2 and R_3 with the center tap connected to the inverting input of the op amp. The cutoff frequency f_c may be adjusted by varying R_1. The amplitude response should look like that of Fig. 2-14(b)

    The first-order low-pass filter stage was discussed in Section 2-9.

3

LOW-PASS FILTERS— INVERSE CHEBYSHEV AND ELLIPTIC TYPES

3-1 INVERSE CHEBYSHEV FILTERS

The *inverse Chebyshev* low-pass filter has the amplitude response [33] given by

$$|H(j\omega)| = \frac{\epsilon C_n(\omega_1/\omega)}{\sqrt{1 + \epsilon^2 C_n^2(\omega_1/\omega)}} \qquad (n = 1, 2, 3, \ldots) \qquad (3\text{-}1)$$

where ϵ is a positive constant and C_n is the Chebyshev polynomial that we associated with the Chebyshev filter in Chapter 2. The constant ω_1 is the beginning frequency of the stopband, as shown in Fig. 3-1 for the case $n = 6$.

The 3-dB cutoff point ω_c, shown also in Fig. 3-1, is given by

$$\omega_c = \frac{\omega_1}{\cosh\left[(1/n)\operatorname{arccosh}(1/\epsilon)\right]} \qquad (3\text{-}2)$$

The response is monotonic in the passband, $0 < \omega < \omega_c$, and has

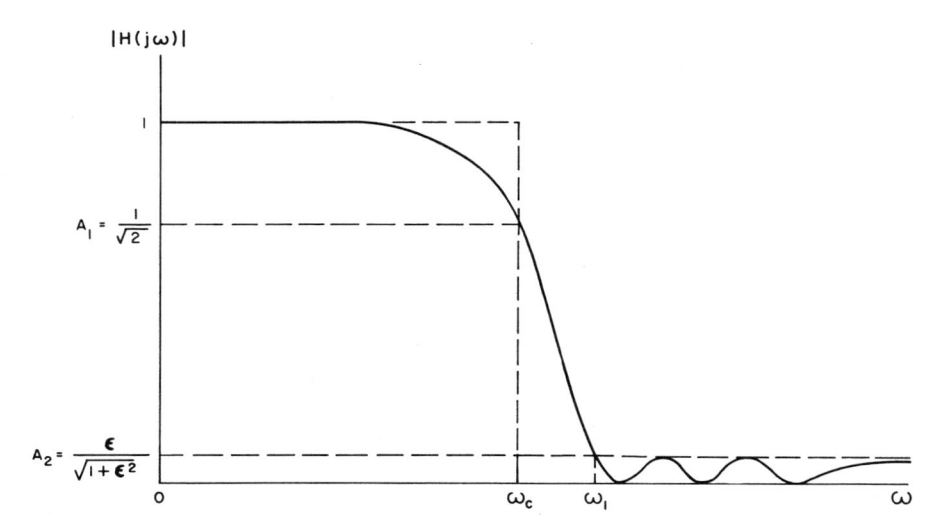

FIGURE 3-1 *Inverse Chebyshev low-pass amplitude response for the case* n = 6.

ripples in the stopband $\omega > \omega_1$ that are equal in magnitude to $A_2 = \epsilon/\sqrt{1+\epsilon^2}$. The transition width is given by

$$\text{TW} = \omega_1 - \omega_c \tag{3-3}$$

If $\alpha_2 = -20 \log_{10} A_2$ is the minimum stopband loss in dB, then ϵ is determined by

$$\epsilon = \frac{1}{\sqrt{10^{\alpha_2/10} - 1}} \tag{3-4}$$

Thus for a given order n, a given minimum allowable stopband loss α_2, and a given ω_1 (the beginning of the stopband ripple channel), we may find ϵ from Eq. (3-4) and the required amplitude response from Eq. (3-1). The cutoff frequency ω_c and the transition width are then found from Eqs. (3-2) and (3-3). Alternatively, we may specify ω_c (rather than ω_1) and find ω_1 from Eq. (3-2).

The amplitude response of a sixth-order inverse Chebyshev filter that was constructed in the laboratory is shown in Fig. 3-2. The minimum stopband loss is 20 dB.

The minimum order n required for a given set of specifications α_2, ω_c, and ω_1 is given from Eqs. (3-2) and (3-4) by

$$n = \frac{\operatorname{arccosh} \sqrt{10^{\alpha_2/10} - 1}}{\operatorname{arccosh} (\omega_1/\omega_c)} \tag{3-5}$$

SEC. 3-1   INVERSE CHEBYSHEV FILTERS   43

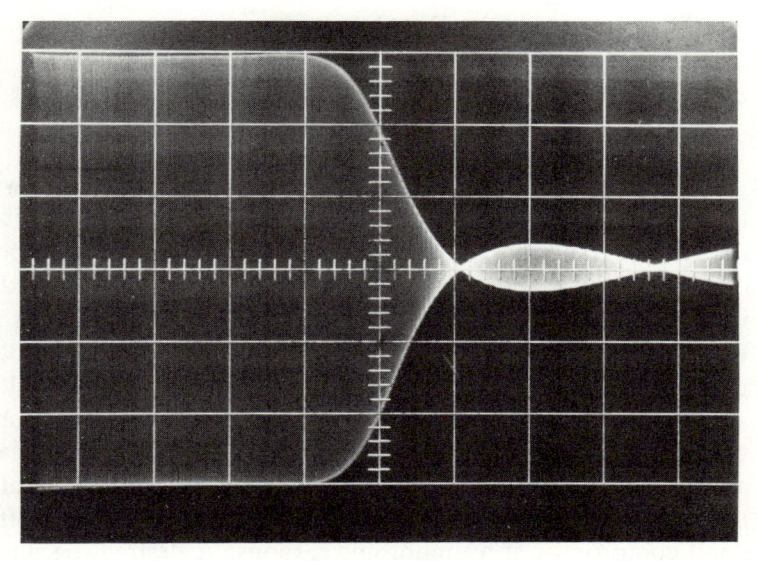

FIGURE 3-2 *Response of an actual inverse Chebyshev filter of order 6.*

The transition width TW (in rad/s) and the ratio ω_1/ω_c are related from Eq. (3-3) by

$$\frac{\omega_1}{\omega_c} = \frac{\text{TW}}{\omega_c} + 1 \qquad (3\text{-}6)$$

Also, by Eqs. (3-5) and (3-6), we have the normalized transition width

$$\frac{\text{TW}}{\omega_c} = \cosh\left(\frac{1}{n} \operatorname{arccosh} \sqrt{10^{\alpha_2/10} - 1}\right) - 1 \qquad (3\text{-}7)$$

Thus we see that a smaller transition width requires a higher value of n, which is accompanied by more ripples.

As an example, suppose that we wish to find the minimum order required for $\alpha_2 = 20$ dB, $f_c = \omega_c/2\pi = 1000$ Hz, and TW ≤ 300 Hz. From Eq. (3-6) we have $\omega_1/\omega_c = 1.3$, so that by Eq. (3-5) the required order is

$$n = \frac{\operatorname{arccosh} \sqrt{10^2 - 1}}{\operatorname{arccosh} 1.3} = 3.95$$

Thus, the minimum order is $n = 4$, which results in TW < 300

Hz. The exact value, by Eq. (3-7), is

$$TW = 1000[\cosh(\tfrac{1}{4} \operatorname{arccosh} \sqrt{99}) - 1]$$
$$= 292$$

This example was considered for the case of a Chebyshev filter in Section 2-4, with identical results. This is an illustration of the general result that for a given allowable passband and stopband deviation and a given ω_c and TW, the Chebyshev filter and the inverse Chebyshev filter require the same order, which is less than the order required for a Butterworth filter. Thus, if a monotonic passband characteristic is required, the inverse Chebyshev filter is superior to a Butterworth filter of the same order. If passband ripples are to be tolerated, the Chebyshev filter is better because, as we shall see, its transfer function is simpler than that of the inverse Chebyshev filter. The Butterworth filter, however, is often a good compromise if a monotonic response is desired, because its transfer function is also simpler than that of the inverse Chebyshev filter.

The transfer function of an inverse Chebyshev filter is of the general form of Eq. (1-6). Thus it is not an all-pole filter and in general is more difficult to realize than the all-pole filters, such as the Butterworth and Chebyshev filters. In factored form the inverse Chebyshev low-pass function is given for n even by

$$H(s) = \prod_{i=1}^{n/2} \frac{A_i(s^2 + a_i)}{s^2 + b_i s + c_i} \tag{3-8}$$

and for n odd by

$$H(s) = \frac{A_0}{s + c_0} \prod_{i=1}^{(n-1)/2} \frac{A_i(s^2 + a_i)}{s^2 + b_i s + c_i} \tag{3-9}$$

where A_0, c_0, A_i, a_i, b_i, and c_i are prescribed constants.

For the convenience of the reader we have tabulated the inverse Chebyshev low-pass transfer functions in Appendix B for the normalized case ($\omega_c = 1$ rad/s). The orders tabulated are for $n = 2, 3, \ldots, 10$ with minimum stopband ripples, in most cases, of 30 to 100 dB in steps of 5 dB. Each factor in Eq. (3-8) or (3-9) corresponds to the filter function if $n = 2$ or to a stage function if $n > 2$. The data for a given transfer function with a specified order n (written as N) and a minimum stopband loss (MSL) are grouped together. For second-order factors the coefficients a_i, b_i, and c_i (indicated as A, B, and C) are listed by lines. The first line corresponds to the first factor, second line to the second factor, and so on. In the case (3-9) of an odd-order filter, the coefficient c_0 is given under C in the last line of the data. The columns WZ, WM, and KM are for use in tuning the filter and will be discussed in Section 3-6.

The constants A_0 and A_i are related to the gain of the filter, or one of its stages. For example, if K_i is the gain of a second-order stage, then K_i is the value of the second-order factor at $s = 0$, given by

$$K_i = \frac{A_i a_i}{c_i}$$

Therefore, we have

$$A_i = \frac{K_i c_i}{a_i} \qquad (3\text{-}10)$$

In the case of the first-order section associated with the first-order factor of Eq. (3-9), its gain, say K_0, is given by $K_0 = A_0/c_0$, and thus we have

$$A_0 = K_0 c_0 \qquad (3\text{-}11)$$

As an example, suppose that we want the transfer function of a fifth-order inverse Chebyshev filter with $\omega_c = 1$ rad/s, a gain of 8, and a minimum stopband loss of MSL = 40 dB. From Appendix B we have for the first second-order stage,

$$A = 2.887037$$
$$B = 0.503909$$
$$C = 1.037939$$

for the second second-order stage,

$$A = 7.558361$$
$$B = 1.696117$$
$$C = 1.334444$$

and for the first-order stage,

$$C = 1.273011$$

3-2 ELLIPTIC FILTERS

As pointed out in Chapter 2, the *elliptic filter* has an amplitude response that exhibits ripples in both the pass- and stopbands, and it is the best of all low-pass filters in that for a given order and allowable pass- and stopband deviations, it has the shortest transition width. An example of a fifth-order elliptic amplitude response is shown in Fig. 3-3.

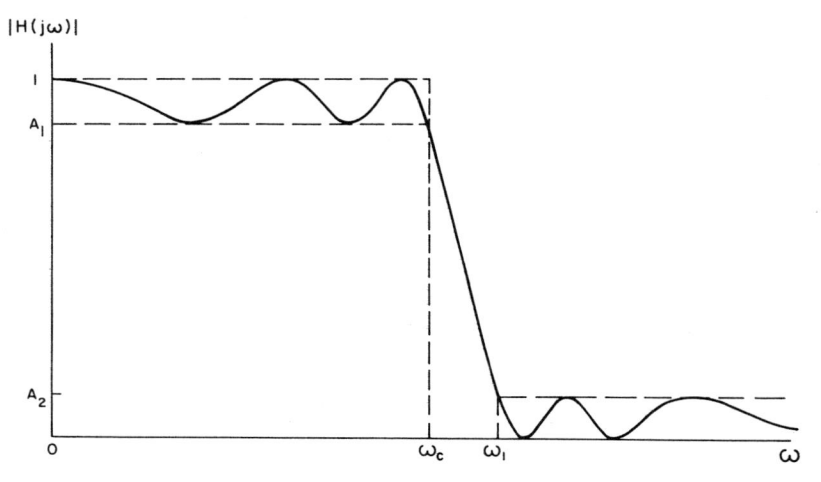

FIGURE 3-3 *Elliptic low-pass amplitude response for the case* n = 5.

The passband ripples are equal in magnitude and may be characterized by the maximum allowable passband loss. This value, which we shall also call the *passband ripple width* (PRW), is given by Fig. 3-3 to be

$$\text{PRW} = -20 \log_{10} A_1 \quad \text{dB} \tag{3-12}$$

The stopband ripples are also equal in magnitude (although not necessarily equal to the passband ripple magnitude) and are characterized by the *minimum stopband loss* (MSL), given by

$$\text{MSL} = -20 \log_{10} A_2 \quad \text{dB} \tag{3-13}$$

The transition width TW is given, as in the other filter types, by

$$\text{TW} = \omega_1 - \omega_c \tag{3-14}$$

SEC. 3-2   ELLIPTIC FILTERS   47

For given values of PRW and MSL, increasing the order increases the number of pass- and stopband ripples and decreases TW. Thus we may specify A_1, A_2, and ω_c in Fig. 3-3, and increase the order to achieve any given $\omega_1 > \omega_c$.

To illustrate the superiority of the elliptic filter, let us consider Fig. 3-4. The two curves shown are plots for the elliptic and Chebyshev filters of order versus transition width. The case considered is that of a 0.1-dB passband ripple and a minimum stopband loss of 60 dB with a cutoff frequency of 1 rad/s. Other cases yield similar results.

As an example of the use of Fig. 3-4, suppose that we want a normalized transition width not to exceed 0.1. In other words, referring to Fig. 3-3 for $\omega_c = 1$, we want ω_1 less than or equal to

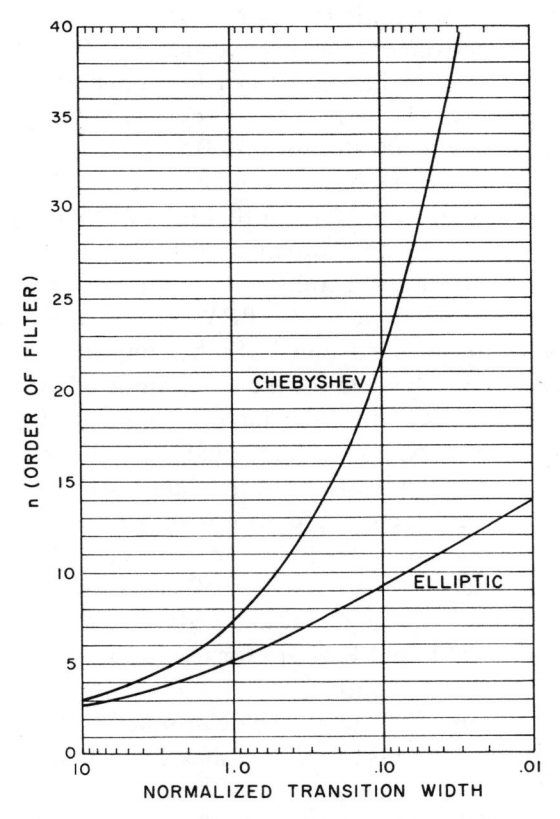

FIGURE 3-4 *A comparison of elliptic and Chebyshev transition widths for PRW = 0.1 and MSL = 60 dB.*

1.1. By Fig. 3-4 we see that an elliptic filter of order 10 is sufficient. (The width 0.1 requires n to be between 9 and 10, and thus we must use $n = 10$.) For the Chebyshev filter, however, the minimum order required is 22. The superiority of the elliptic over the Chebyshev filter is even more remarkable for smaller transition widths. For example, if the width is not to exceed 0.03, an elliptic filter of order 12 is sufficient, but the minimum Chebyshev order is 39.

The transfer function of an elliptic filter has the identical form of that of the inverse Chebyshev filter given previously in Eqs. (3-8) and (3-9) [16]. The constants a_i, b_i, and c_i, which are different, of course, from those of the inverse Chebyshev filter, are extremely tedious to calculate. The process requires a knowledge of the Jacobi elliptic functions (see, for example, [11]).

For the convenience of the reader, we have tabulated the transfer functions in factored form in Appendix C for the normalized case ($\omega_c = 1$ rad/s), for orders $n = 2, 3, \ldots, 10$. The passband ripples available are 0.1, 0.5, 1, 2, and 3 dB, and the minimum stopband ripples are, in most cases, 30 to 100 dB in steps of 5 dB. The transition width which results is given in each case and is, of course, a normalized value, $\omega_1 - 1$.

The data for a given transfer function with a specified order $n = N$, a passband ripple width (PRW), and a minimum stopband loss (MSL), are grouped under the heading of N and PRW. In this group we find the data for the given value of MSL listed in the first column. For second-order factors such as those of Eq. (3-8), the coefficients a_i, b_i, and c_i (indicated as A, B, and C) are listed by lines. The first line corresponds to the first factor, the second line to the second factor, and so on. In the case of a first-order factor (for N odd) such as $A_0/(s + c_0)$ in Eq. (3-9), the coefficient c_0 is given under C in the last line of the data. The column TW denotes the normalized transition width which results, and the columns WZ, WM, and KM are for use in tuning the filter. These will be discussed in Section 3-6.

The constants A_0 and A_i are determined by the gain desired, as in the case of the inverse Chebyshev filter described in Section 3-1.

Alternatively, we may wish to find the transfer function of minimum order N for a given PRW, MSL, and a maximum allow-

SEC. 3-2　　　　　ELLIPTIC FILTERS　　　　　　　　**49**

able transition width TW. This may be done by means of Appendix D, where TW is listed for each elliptic filter case. We need only refer to the appropriate case (that of the given PRW and MSL) and find the smallest N corresponding to a transition width which does not exceed the given TW. The values of TW are normalized ($\omega_c = 1$ rad/s) in both Appendices C and D, and must be multiplied by the given ω_c. In other words, the transition widths which are listed are actually TW/ω_c if TW is in rad/s, or TW/f_c if TW is in Hz. Since these quantities are dimensionless, the values in the Appendices are independent of whether Hz or rad/s is used for the transition width.

As an example, suppose that we want the minimum order N for a cutoff frequency of $f_c = 1000$ Hz, PRW $= 0.1$ dB, MSL $= 30$ dB, and TW ≤ 100 Hz. The normalized TW is then $100/1000 = 0.1$. Locating PRW $= 0.1$ and MSL $= 30$ in Appendix D, we find that $N = 6$ yields TW $= 0.1025$ and $N = 7$ yields TW $= 0.0479$. Therefore, the minimum order is 7.

The amplitude response of an actual elliptic filter is shown in Fig. 3-5. It is a fifth-order example with PRW $= 0.5$ and MSL $= 35$ dB.

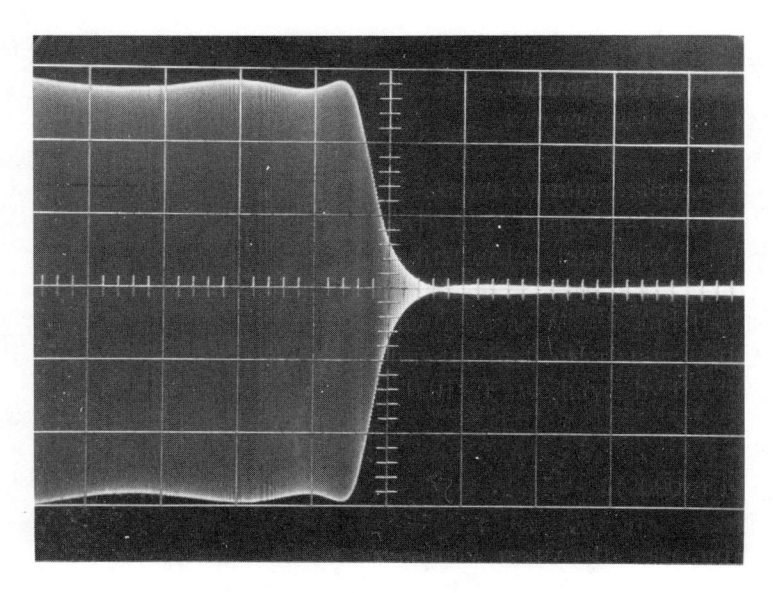

FIGURE 3-5 Response of an actual fifth-order elliptic low-pass filter.

3-3 VCVS ELLIPTIC FILTER CIRCUITS

The inverse Chebyshev and elliptic low-pass filters have transfer functions that are identical in form. In the case of a second-order filter, or a second-order stage of a higher-order filter, with cutoff frequency ω_c rad/s or $f_c = \omega_c/2\pi$ Hz, and gain K, the transfer function is of the form

$$\frac{V_2}{V_1} = \frac{KC}{A}\left[\frac{s^2 + A\omega_c^2}{s^2 + B\omega_c s + C\omega_c^2}\right] \quad (3\text{-}15)$$

The coefficients A, B, and C may be found in Appendix B in the case of the inverse Chebyshev filter and in Appendix C in the case of the elliptic filter. They depend on the order N, the minimum stopband loss MSL, and, in the case of the elliptic filter, the passband ripple width PRW.

Equation (3-15) is of the general form

$$\frac{V_2}{V_1} = \frac{\rho(s^2 + \alpha\omega_c^2)}{s^2 + \beta\omega_c s + \gamma\omega_c^2} \quad (3\text{-}16)$$

where

$$\rho = \frac{KC}{A}, \quad \alpha = A, \quad \beta = B, \quad \gamma = C \quad (3\text{-}17)$$

As we shall see later, Eq. (3-16) is also the form of second-order elliptic and inverse Chebyshev filter functions of high-pass, band-pass, and band-reject types.

There are a number of circuits which realize the second-order function (3-16). One of the simplest is the VCVS circuit of Fig. 3-6 [23], for which

$$\rho = -\frac{R_4}{R_5}$$

$$\alpha\omega_c^2 = \frac{R_5}{R_1 R_2 R_4 C_1 C_2}$$

$$\beta\omega_c = \frac{1}{R_2 C_2} \quad (3\text{-}18)$$

$$\gamma\omega_c^2 = \frac{1}{R_2 R_3 C_1 C_2}$$

SEC. 3-3   VCVS ELLIPTIC FILTER CIRCUITS   51

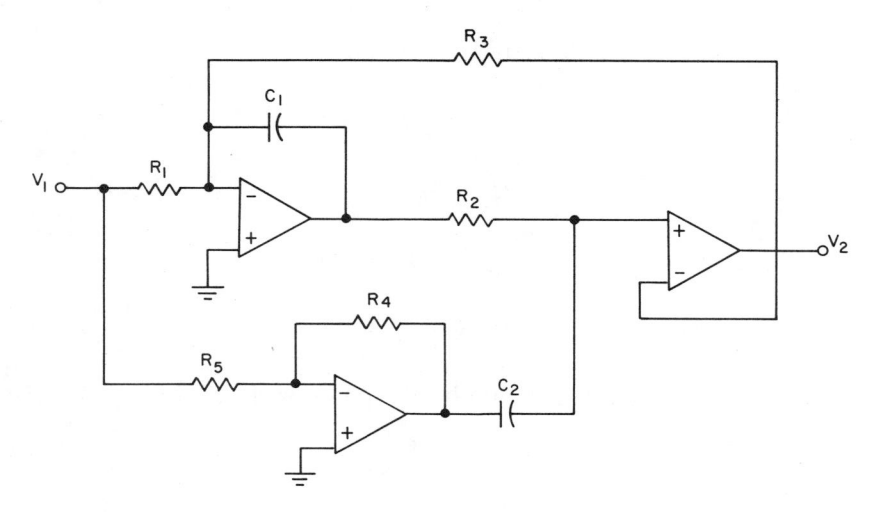

FIGURE 3-6 *Second-order low-pass voltage follower elliptic filter.*

Solving for the resistances and replacing ρ, α, β, and γ by their values in Eq. (3-17), we have

$$R_1 = -\frac{\beta}{\alpha\rho\omega_c C_1} = \frac{\beta}{KC\omega_c C_1}$$

$$R_2 = \frac{1}{\beta\omega_c C_2} = \frac{1}{B\omega_c C_2}$$

$$R_3 = \frac{\beta}{\gamma\omega_c C_1} = KR_1 \qquad (3\text{-}19)$$

$$R_4 = -\rho R_5 = \frac{KCR_5}{A}$$

where R_5, C_1, and C_2 are arbitrary. The gain is an inverting gain $-K$ ($K > 0$). To distinguish this circuit from others we shall consider, we will refer to it as a *voltage-follower circuit* because one of the op amps is operated as a voltage follower, as discussed in Section 2-6.

If $C_1 = C_2$ is chosen as a standard value near $10/f_c$ μF, a reasonable choice of R_5 is

$$R_5 = \frac{1}{\omega_c C_1} \qquad (3\text{-}20)$$

in which case the other resistances are

$$R_1 = \frac{BR_5}{KC}$$

$$R_2 = \frac{R_5}{B}$$

$$R_3 = \frac{BR_5}{C} = KR_1 \qquad (3\text{-}21)$$

$$R_4 = \frac{KCR_5}{A}$$

A variation of the circuit of Fig. 3-6 is shown in Fig. 3-7. To

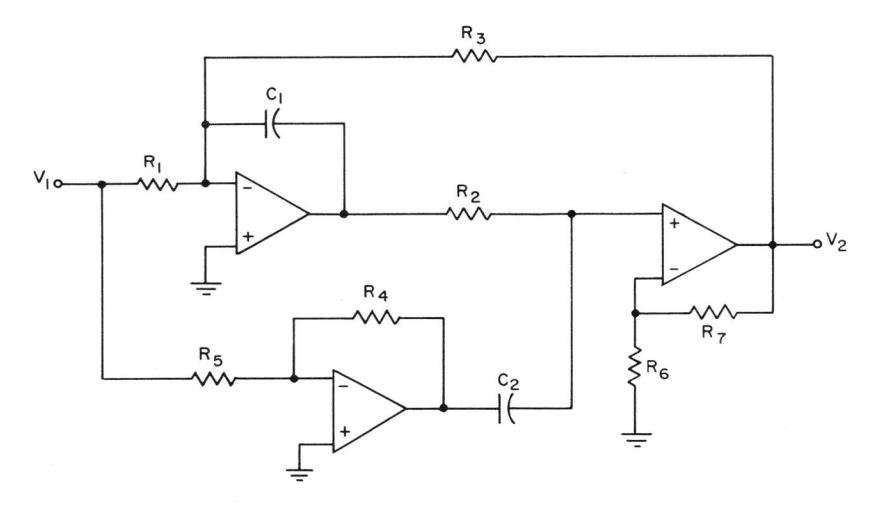

FIGURE 3-7 *Second-order low-pass VCVS elliptic filter.*

distinguish this circuit from that of Fig. 3-6, we shall refer to it as a VCVS circuit because of the mode in which one of the op amps is operating. The function of Eq. (3-16) is realized by Fig. 3-7 with

$$\rho = -\frac{\mu R_4}{R_5}$$

$$\alpha \omega_c^2 = \frac{R_5}{R_1 R_2 R_4 C_1 C_2}$$

$$\beta \omega_c = \frac{1}{R_2 C_2} \qquad (3\text{-}22)$$

$$\gamma \omega_c^2 = \frac{\mu}{R_2 R_3 C_1 C_2}$$

SEC. 3-3   VCVS ELLIPTIC FILTER CIRCUITS   53

and
$$\mu = 1 + \frac{R_7}{R_6}$$

is the gain of the VCVS. Solving for the resistances and replacing ρ, α, β, and γ by their values in Eq. (3-17), we have

$$
\begin{aligned}
R_1 &= -\frac{\mu\beta}{\rho\alpha\omega_c C_1} = \frac{\mu B}{KC\omega_c C_1} \\
R_2 &= \frac{1}{\beta\omega_c C_2} = \frac{1}{B\omega_c C_2} \\
R_3 &= \frac{\mu\beta}{\gamma\omega_c C_1} = KR_1 \\
R_4 &= -\frac{\rho R_5}{\mu} = \frac{KCR_5}{\mu A} \\
R_6 &= \frac{\mu R_2}{\mu - 1} \quad (\mu \neq 1) \\
R_7 &= \mu R_2
\end{aligned}
\qquad (3\text{-}23)
$$

where C_1, C_2, $\mu > 1$, and R_5 are arbitrary. If we desire $\mu = 1$, the circuit becomes that of Fig. 3-6.

If in Fig. 3-7 we choose $C_1 = C_2$ near $10/f_c$ μF, a reasonable value of R_5 is

$$R_5 = \frac{1}{\omega_c C_1} \qquad (3\text{-}24)$$

in which case the other resistances are

$$
\begin{aligned}
R_1 &= \frac{\mu B R_5}{KC} \\
R_2 &= \frac{R_5}{B} \\
R_3 &= \frac{\mu B R_5}{C} = KR_1 \\
R_4 &= \frac{KCR_5}{\mu A} \\
R_6 &= \frac{\mu R_5}{B(\mu - 1)} \\
R_7 &= \frac{\mu R_5}{B}
\end{aligned}
\qquad (3\text{-}25)
$$

If K and the pole-pair Q (given by $\sqrt{C/B}$) are moderate values, the resistances in Eqs. (3-20) and (3-21) for Fig. 3-6 and Eqs. (3-24) and (3-25) for Fig. 3-7 will be reasonable values. However, if Q and/or K are high, say over 10, there may be an undesirable spread of resistance values. In this case Eq. (3-19) or (3-23) may be used and C_1, C_2, and R_s may be chosen to maintain a lower spread of resistance values. In the case of Eq. (3-23), μ is also available as a variable parameter. For example, if Q is high (B is low), we may make C_2 relatively large compared to C_1 to keep R_2 in somewhat the same range as R_1 and R_3.

The design procedure for the VCVS filter is summarized in Section 3-8.

3-4 ELLIPTIC FILTER CIRCUITS WITH THREE CAPACITORS

Another example of a second-order low-pass filter is the three-capacitor circuit of Fig. 3-8 [1], which realizes Eq. (3-16) with

$$\rho = -\frac{C_3}{C_2}$$

$$\alpha \omega_c^2 = \frac{1}{R_1 R_4 C_1 C_3}$$

$$\beta \omega_c = \frac{1}{R_3 C_2} \qquad (3\text{-}26)$$

$$\gamma \omega_c^2 = \frac{1}{R_2 R_4 C_1 C_2}$$

Solving for the element values and replacing ρ, α, β, and γ by their values in Eq. (3-17), we have

$$C_3 = -\rho C_2 = \frac{KCC_2}{A}$$

$$R_1 = -\frac{1}{R_4 \rho \alpha \omega_c^2 C_1 C_2} = \frac{1}{R_4 KC \omega_c^2 C_1 C_2}$$

$$R_2 = \frac{1}{R_4 \gamma \omega_c^2 C_1 C_2} = KR_1 \qquad (3\text{-}27)$$

$$R_3 = \frac{1}{\beta \omega_c C_2} = \frac{1}{B \omega_c C_2}$$

SEC. 3-4   ELLIPTIC FILTER CIRCUITS

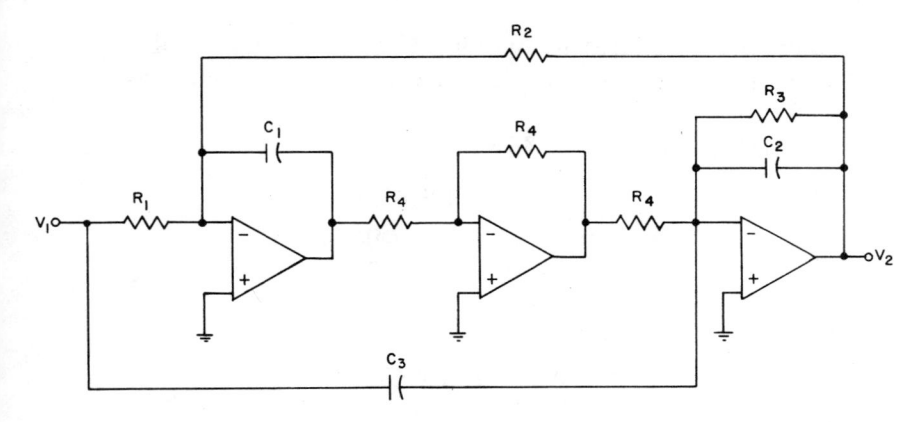

FIGURE 3-8 Second-order low-pass elliptic filter using three capacitors.

where C_1, C_2, and R_4 are arbitrary. The gain is an inverting gain of $-K$ ($K > 0$).

If we choose C_1 arbitrarily (preferably near $10/f_c$ μF), then C_2 may be chosen to give a reasonable value of R_3. If Q is high (B is low), C_2 may be larger, and if Q is low (B is high), C_2 may be smaller. The value of R_4 then may be selected to give reasonable values of R_1 and R_2.

As an example, suppose that we want an eighth-order low-pass elliptic filter with a gain of 16, $f_c = 1000$ Hz, PRW = 0.5 dB, and MSL = 60 dB. There will be four second-order stages with transfer functions like Eq. (3-15), and we shall take the gain per stage to be $K = 2$. We shall give the details for stages 1 and 4, which have, respectively, a low Q of 0.702 and a high Q of 27.481.

By Appendix C we have for stage 1,

$$A = 1.285297$$
$$B = 0.603927$$
$$C = 0.179641$$

Choosing $C_1 = 10/f_c = 0.01$ μF in Fig. 3-8, we have, by Eq. (3-27),

$$C_3 = 0.2795 C_2$$
$$R_1 = \frac{7.0511}{R_4 C_2}$$
$$R_2 = \frac{14.1022}{R_4 C_2}$$
$$R_3 = \frac{263.5334 \times 10^{-6}}{C_2}$$

where the resistances are in ohms. If we choose $C_2 = 0.1\ \mu\text{F}$, then $R_3 = 2.635\ k\Omega$, a reasonable value. If R_4 is then chosen as $10\ k\Omega$, we have $R_1 = 7.051\ k\Omega$ and $R_2 = 14.102\ k\Omega$.

In the case of stage 4, we have

$$A = 1.514535$$
$$B = 0.036505$$
$$C = 1.006426$$

for which, choosing $C_1 = 0.01\ \mu\text{F}$, we have

$$C_3 = 1.3290 C_2$$
$$R_1 = \frac{1.2585}{R_4 C_2}$$
$$R_2 = \frac{2.5170}{R_4 C_2}$$
$$R_3 = \frac{4.3598 \times 10^{-3}}{C_2}$$

Selecting $C_2 = 0.1\ \mu\text{F}$, we have $R_3 = 43.598\ k\Omega$. Finally, choosing $R_4 = 5\ k\Omega$, we have $R_1 = 2.517\ k\Omega$ and $R_2 = 5.034\ k\Omega$.

The other stages are developed in a similar manner and all four stages are then cascaded to form the filter. The actual amplitude response is shown in Fig. 3-9.

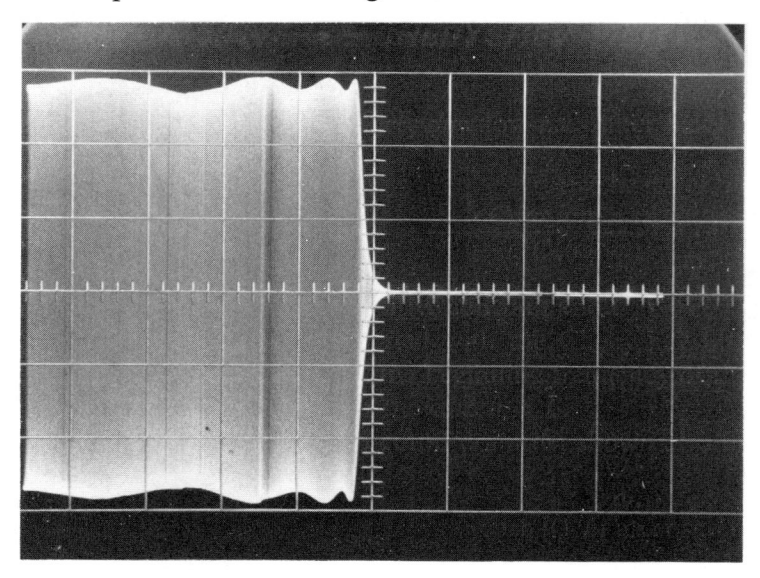

FIGURE 3-9 *Actual response of an eighth-order low-pass elliptic filter.*

SEC. 3-5 BIQUAD ELLIPTIC FILTER CIRCUITS

Figure 3-8 is a more complicated circuit than either Fig. 3-6 or 3-7, requiring three capacitors instead of two. This disadvantage, however, is offset by the ease of tuning that Fig. 3-8 affords, as we shall see in Section 3-6. We note also from the first of Eq. (3-27) that C_3 is not necessarily a standard value and may have to be trimmed, or we may have to take a gain slightly different from K.

The design procedure for the three-capacitor circuit is summarized in Section 3-9.

3-5 BIQUAD ELLIPTIC FILTER CIRCUITS

A final example of a second-order low-pass elliptic filter is the biquad circuit [8] of Fig. 3-10, which we shall call a *biquad elliptic filter circuit*. It achieves Eq. (3-16) with

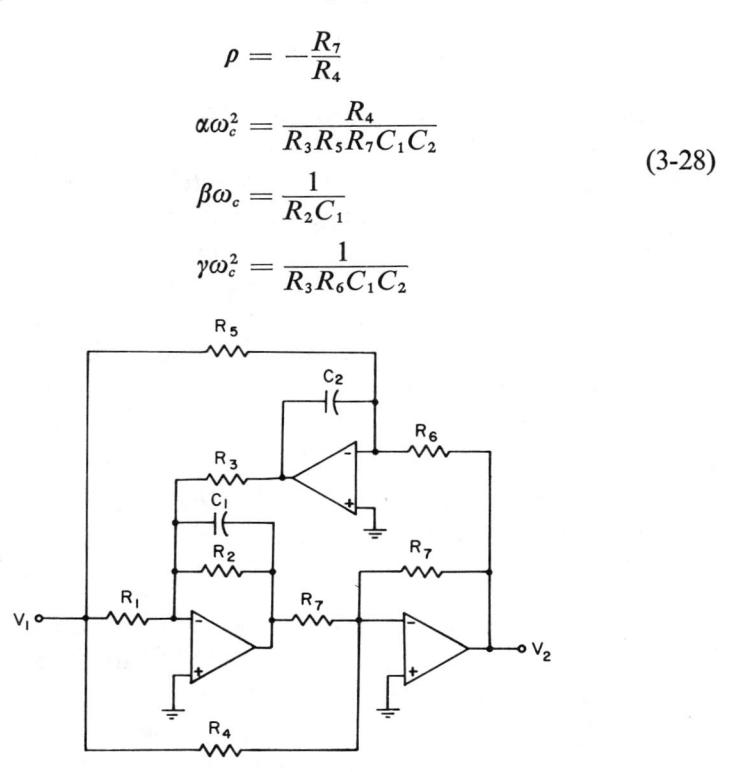

$$\rho = -\frac{R_7}{R_4}$$
$$\alpha\omega_c^2 = \frac{R_4}{R_3 R_5 R_7 C_1 C_2}$$
$$\beta\omega_c = \frac{1}{R_2 C_1} \tag{3-28}$$
$$\gamma\omega_c^2 = \frac{1}{R_3 R_6 C_1 C_2}$$

FIGURE 3-10 Biquad elliptic low-pass filter.

provided that

$$R_2 R_4 = R_1 R_7 \qquad (3\text{-}29)$$

Solving for the resistances and replacing ρ, α, β, and γ by their values in Eq. (3-17), we have [19]

$$\begin{aligned}
R_1 &= -\frac{1}{\rho\beta\omega_c C_1} = \frac{A}{KBC\omega_c C_1} \\
R_2 &= -\rho R_1 = \frac{1}{B\omega_c C_1} \\
R_3 &= \frac{1}{\sqrt{\gamma}\,\omega_c C_1} = \frac{1}{\sqrt{C}\,\omega_c C_1} \\
R_4 &= -\frac{R_7}{\rho} = \frac{AR_7}{KC} \\
R_5 &= -\frac{\sqrt{\gamma}}{\rho\alpha\omega_c C_2} = \frac{1}{K\sqrt{C}\,\omega_c C_2} \\
R_6 &= \frac{C_1}{C_2} R_3
\end{aligned} \qquad (3\text{-}30)$$

where C_1, C_2, and R_7 are arbitrary. The gain is an inverting gain of $-K$ ($K > 0$).

The circuit of Fig. 3-10 is similar in complexity to those of Figs. 3-7 and 3-8. It is easier to tune than the VCVS circuit and has the advantage over the three-capacitor circuit that the gain may be set without the necessity of trimming a capacitor. In all three cases, values of pole-pair Q up to 100 can be achieved.

The design procedure for the biquad elliptic filter is summarized in Section 3-10.

3-6 TUNING THE INVERSE CHEBYSHEV AND ELLIPTIC FILTERS

The tuning of the second-order stages of inverse Chebyshev and elliptic low-pass filters is more readily accomplished by observing the general shape that the amplitude response should take. This is shown in Fig. 3-11(a) in the case of amplitude peaking in the passband, and in Fig. 3-11(b) if there is no passband peaking. In the case of Fig. 3-11(a), the peak value K_m and the frequency f_m

SEC. 3-6 TUNING THE ELLIPTIC FILTERS

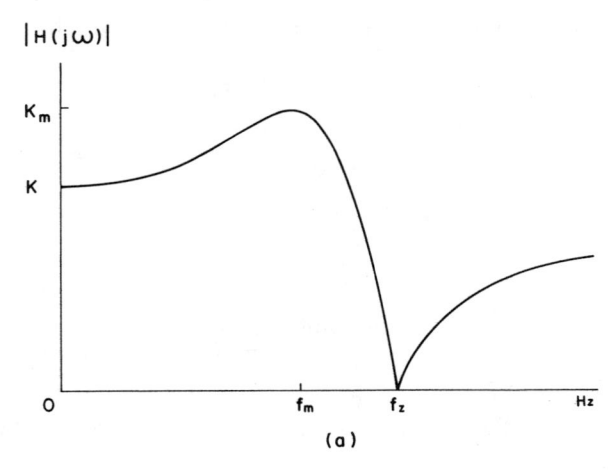

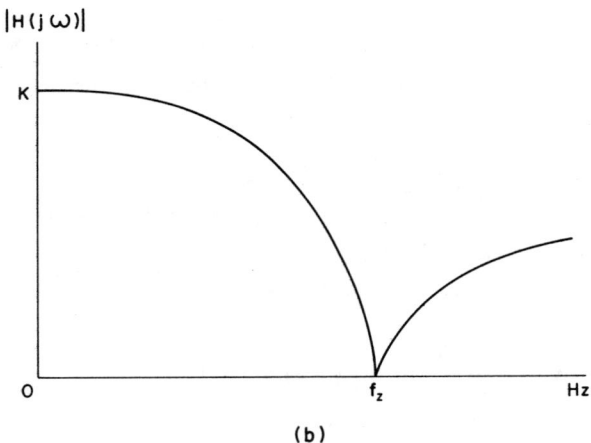

FIGURE 3-11 Low-pass inverse Chebyshev or elliptic responses with: (a) passband peaking and (b) without passband peaking.

(Hz) at which it occurs are given by

$$K_m = \frac{2KC}{AB}\sqrt{\frac{(A-C)^2 + AB^2}{4C - B^2}} \qquad (3\text{-}31)$$

and

$$f_m = f_c\sqrt{\frac{2C(A-C) - AB^2}{2(A-C) + B^2}} \qquad (3\text{-}32)$$

There is no passband peaking, of course, if f_m in Eq. (3-32) is an imaginary number.

In both figures the notch frequency f_z, at which the amplitude is zero, is given by

$$f_z = f_c \sqrt{A} \qquad (3\text{-}33)$$

and the value K_c of the amplitude at the cutoff frequency f_c is given by

$$K_c = \frac{KC}{A} \frac{|A-1|}{\sqrt{(C-1)^2 + B^2}} \qquad (3\text{-}34)$$

The values of ω_z, ω_m, and K_m/K are given in Appendices B and C for each filter stage as WZ, WM, and KM, respectively. The values of ω_z and ω_m are normalized to $\omega_c = 1$.

To tune the circuit of Fig. 3-7, we may perform the following steps:

1. Adjust the ratio R_4/R_5 slightly to move the notch toward f_z
2. Adjust $\mu = 1 + R_7/R_6$ slightly to move the peak toward f_m
3. Adjust R_2 slightly to obtain K_m

These steps may be repeated if necessary.

The three-capacitor circuit of Fig. 3-8 is easier to tune. The following steps performed once usually will be sufficient, except for low-Q sections, in which case steps 2 and 3 may have to be repeated:

1. Adjust R_1 until the notch is at f_z
2. Adjust R_2 until the peak is at f_m
3. Adjust R_3 until K_m is attained

In the case of the biquad circuit of Fig. 3-10, the notch can be set by adjusting R_4, the cutoff frequency $\sqrt{C}\,\omega_c$ by R_3, the pole-pair Q, related to $B\omega_c$, by R_2, and the gain by R_1.

3-7 ODD-ORDER ELLIPTIC FILTERS

In the case of odd-order elliptic or inverse Chebyshev filters, one stage must have a first-order transfer function

$$\frac{V_2}{V_1} = \frac{KC\omega_c}{s + C\omega_c} \qquad (3\text{-}35)$$

SEC. 3-8 VCVS LOW-PASS ELLIPTIC FILTER DESIGN

where K is the stage gain, ω_c the cutoff frequency of the filter, and C a constant found in the last line of the transfer function data for the filter in Appendices B and C.

Equation (3-35) is identical to the first-order function for the Butterworth and Chebyshev filters discussed in Section 2-9. Therefore, we may realize Eq. (3-35) by means of the circuits of Fig. 2-15 or 2-16. The design procedure is summarized in Section 3-11.

3-8 VCVS LOW-PASS ELLIPTIC FILTER DESIGN SUMMARY

To design a second-order low-pass elliptic or inverse Chebyshev filter, or a second-order stage of a higher-order filter, having a given cutoff frequency f_c Hz (or $\omega_c = 2\pi f_c$ rad/s), gain K, minimum stopband loss (MSL), and in the case of the elliptic filter, passband ripple width (PRW), perform the following steps.

1. Find the normalized coefficients A, B, and C from the appropriate table of Appendix B or C
2. Select a standard value of C_1 (preferably near $10/f_c$ μF) and calculate the resistance values given by

$$R_1 = \frac{\mu B}{KC\omega_c C_1}$$

$$R_2 = \frac{1}{B\omega_c C_2}$$

$$R_3 = KR_1$$

$$R_4 = \frac{KCR_5}{\mu A}$$

$$R_6 = \frac{\mu R_2}{\mu - 1}$$

$$R_7 = \mu R_2$$

where C_2, R_5, and $\mu > 1$ are arbitrary. If K and the pole-pair $Q = \sqrt{C}/B$ are moderate values, such as 10 or less, reasonable values of the arbitrary parameters are

$$C_2 = C_1$$

$$R_5 = \frac{1}{\omega_c C_1}$$

and $\mu = 2$ (in which case $R_6 = R_7$). If Q and/or K are high, say over 10, C_2, R_5, and μ should be chosen to maintain a lower spread of resistance values. For example, if Q is high (B is low), C_2 may be made relatively large compared to C_1 to keep R_2 in somewhat the same range as R_1 and R_3

3. Select standard values of resistances as close as possible to the calculated values and construct the filter, or its stages, in accordance with Fig. 3-12

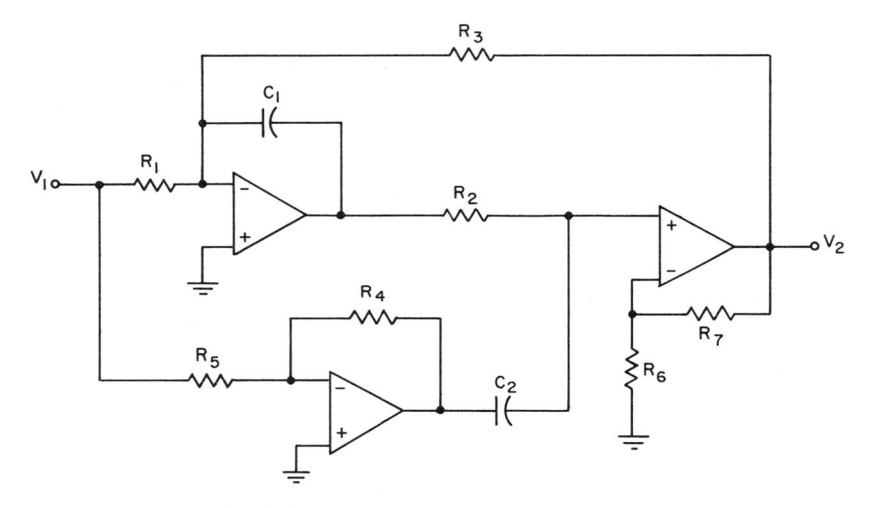

FIGURE 3-12 *VCVS low-pass elliptic filter circuit.*

4. If $\mu = 1$ is desired, R_6 becomes an open circuit and R_7 becomes a short circuit, with the other resistances given as in step 2. In this case the circuit is the voltage-follower circuit of Fig. 3-6

Comments

(a) Comments (a) and (b) for the MFB filter of Section 2-10 apply directly, except that in (b), R_{eq} for each op amp is the resistance R_1, R_2, or R_5 connected to its input terminal

(b) The circuit may be used for both high and low pole-pair Q by selecting the arbitrary parameters of step 2 to keep the spread of resistance values relatively low. For moderate gains, Q may range as high as 100

(c) Tuning may be accomplished by varying the ratio R_4/R_5 to move the notch toward f_z, shown in Fig. 3-11. The gain μ

SEC. 3-9   3-CAPACITOR LP ELLIPTIC FILTER DESIGN

of the VCVS, given by

$$\mu = 1 + \frac{R_7}{R_6}$$

may be adjusted, by varying the ratio R_7/R_6, to move the peak toward f_m. Finally, R_2 may be varied to obtain K_m. These steps may be repeated until the stage is tuned

(d) The resulting transition width is TW $\times f_c$ Hz, where TW is the normalized transition width given in Appendix B or C. Alternatively, one may desire a minimum transition width and find the required filter order from Eqs. (3-5) and (3-6) in the case of the inverse Chebyshev filter, or from Appendix D in the case of the elliptic filter

(e) The circuit yields an inverting gain with magnitude $K = R_3/R_1$

The VCVS elliptic filter was discussed in Section 3-3.

3-9 THREE-CAPACITOR LOW-PASS ELLIPTIC FILTER DESIGN SUMMARY

To design a second-order low-pass elliptic or inverse Chebyshev filter, or a second-order stage of a higher-order filter, having a given cutoff frequency f_c Hz (or $\omega_c = 2\pi f_c$ rad/s), gain K, minimum stopband loss (MSL), and in the case of the elliptic filter, passband ripple width (PRW), perform the following steps.

1. Find the normalized coefficients A, B, and C from the appropriate table of Appendix B or C

2. Select a standard value of C_1 (preferably near $10/f_c$ μF) and calculate the element values given by

$$C_3 = \frac{KCC_2}{A}$$

$$R_1 = \frac{1}{R_4 A \omega_c^2 C_1 C_3}$$

$$R_2 = KR_1$$

$$R_3 = \frac{1}{B\omega_c C_2}$$

where C_2 and R_4 are arbitrary. If the pole-pair $Q = \sqrt{C}/B$ is high, C_2 may be chosen near C_1, and if Q is low, C_2 may be chosen larger than C_1. In either case, R_4 may be selected to minimize the spread of the resistance values

3. Select standard values of resistances and capacitances as close as possible to the calculated values and construct the filter, or its stages, in accordance with Fig. 3-13

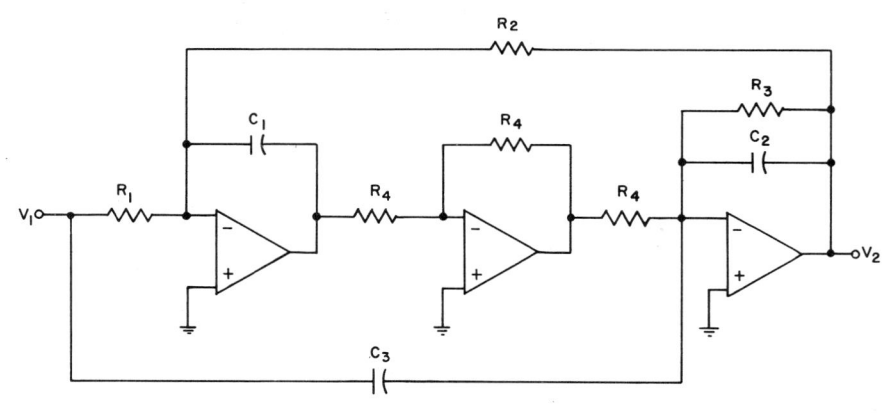

FIGURE 3-13 *Three-capacitor low-pass elliptic filter circuit.*

Comments

(a) Comments (a) and (b) for the MFB filter of Section 2-10 apply directly, except that in (b), R_{eq} for each op amp is the resistance R_1 or R_4 connected to its inverting input

(b) The circuit may be used for both low and high pole-pair Q, with an upper limit of approximately $Q = 100$

(c) Tuning may be accomplished as discussed in Section 3-6 by first adjusting R_1 until the notch is at f_z, then adjusting R_2 to place the peak amplitude at f_m, and finally adjusting R_3 to obtain the correct value of K_m. For low-Q sections, the last two steps may need to be repeated

(d) Comment (d) for the VCVS elliptic filter of Section 3-8 applies directly

(e) The circuit yields an inverting gain with magnitude $K = R_2/R_1$

SEC. 3-10   BIQUAD LP ELLIPTIC FILTER DESIGN   65

The three-capacitor circuit was discussed and an example given in Section 3-4.

3-10 BIQUAD LOW-PASS ELLIPTIC FILTER DESIGN SUMMARY

To design a second-order low-pass elliptic or inverse Chebyshev filter, or a second-order stage of a higher-order filter, having a given cutoff frequency f_c Hz (or $\omega_c = 2\pi f_c$ rad/s), gain K, minimum stopband loss (MSL), and in the case of the elliptic filter, passband ripple width (PRW), perform the following steps.

1. Find the normalized coefficients A, B, and C from the appropriate table of Appendix B or C

2. Select a standard value of C_1 (preferably near $10/f_c$ μF) and calculate the element values given by

$$R_1 = \frac{A}{KBC\omega_c C_1}$$

$$R_2 = \frac{1}{B\omega_c C_1}$$

$$R_3 = \frac{1}{\sqrt{C}\,\omega_c C_1}$$

$$R_4 = \frac{AR_7}{KC}$$

$$R_5 = \frac{1}{K\sqrt{C}\,\omega_c C_2}$$

$$R_6 = \frac{C_1 R_3}{C_2}$$

The values of C_2 and R_7 are arbitrary, and may be chosen, depending on the gain K and the pole-pair $Q = \sqrt{C}/B$, to minimize the spread of the resistance values. For moderate K and Q, reasonable values are $C_2 = C_1$ and $R_7 = 1/\omega_c C_1$.

3. Select standard values of resistances and capacitances as close as possible to the calculated values and construct the filter, or its stages, in accordance with Fig. 3-14

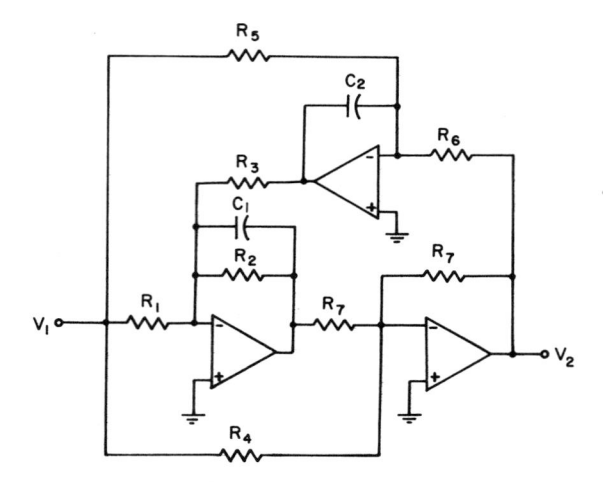

FIGURE 3-14 Biquad low-pass elliptic filter circuit.

Comments

(a) Comments (a) and (b) for the MFB filter of Section 2-10 apply directly, except that in (b), R_{eq} for each op amp is the resistance R_1, R_6, or R_7 connected to its inverting input

(b) The circuit may be used for both low and high pole-pair Q, with an upper limit of approximately $Q = 100$

(c) Tuning may be accomplished by adjusting R_4 to set the notch at f_z, adjusting R_3 to set the cutoff frequency, adjusting R_2 to set Q, and adjusting R_1 or R_5 to set the gain

(d) Comment (d) for the VCVS elliptic filter of Section 3-8 applies directly

(e) The circuit yields an inverting gain with magnitude R_6/R_5.

The biquad elliptic filter was discussed in section 3-5.

3-11 ODD-ORDER LOW-PASS ELLIPTIC FILTER DESIGN SUMMARY

To design a first-order stage of a higher odd-order inverse Chebyshev or elliptic low-pass filter having a given cutoff frequency f_c Hz (or $\omega_c = 2\pi f_c$ rad/s), stage gain K, minimum stop-

SEC. 3-11 ODD-ORDER LP ELLIPTIC FILTER DESIGN

band loss (MSL), and in the case of the elliptic filter, passband ripple width (PRW), perform the following steps.

1. Find the normalized coefficient C for the first-order stage from the appropriate table of Appendix C or D

2. Select a standard value of capacitance C_1 (preferably near $10/f_c\ \mu\text{F}$)

3. (a) If $K > 1$, use the circuit of Fig. 3-15(a) with resistance

(a)

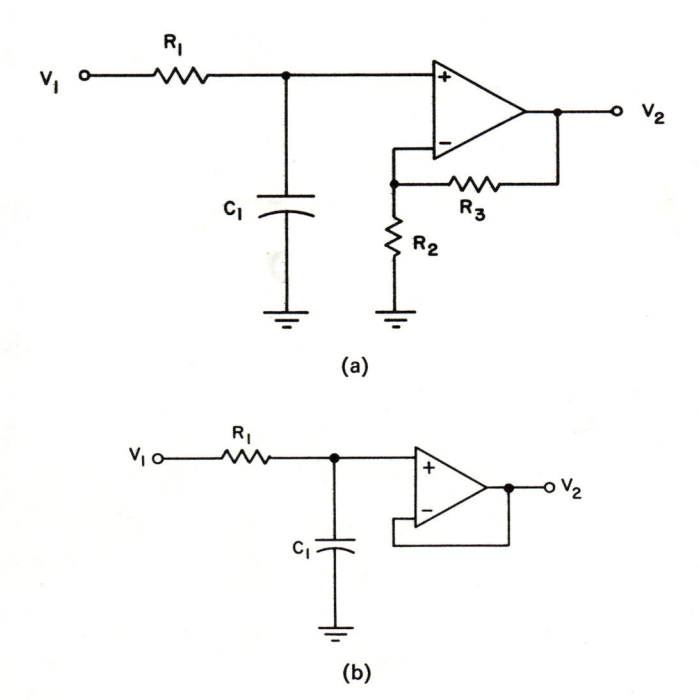

(b)

FIGURE 3-15 First-order low-pass filter circuits.

values given by

$$R_1 = \frac{1}{\omega_c C_1 C}$$

$$R_2 = \frac{KR_1}{K-1}$$

$$R_3 = KR_1$$

(b) If $K = 1$, use the circuit of Fig. 3-15(b) with R_1 as given in step 3(a)
4. The second-order stages of the odd-order filter may be constructed as indicated in Section 3-8, 3-9, or 3-10 and cascaded with the first-order stage to form the filter

Comments

The comments are identical to those of the first-order low-pass design of Section 2-13.

4

HIGH-PASS FILTERS

4-1 THE GENERAL CASE

A high-pass filter is one that passes high-frequency signals and blocks those with low frequencies. Ideal and practical high-pass amplitude responses are shown in Fig. 4-1, where in the practical case the passband is $\omega \geq \omega_c$, the stopband is $0 \leq \omega \leq \omega_1$, the transition band is $\omega_1 < \omega < \omega_c$, and the cutoff frequency is ω_c rad/s, or $f_c = \omega_c/2\pi$ Hz.

The transfer function of a high-pass filter with a cutoff frequency of ω_c rad/s may be obtained from that of a normalized low-pass filter (having a cutoff frequency of 1 rad/s) by replacing s by ω_c/s [32]. A *Butterworth* or *Chebyshev* high-pass filter function, therefore, will contain such second-order factors as

$$H = \frac{V_2}{V_1} = \frac{Ks^2}{s^2 + (B\omega_c/C)s + \omega_c^2/C} \qquad (4\text{-}1)$$

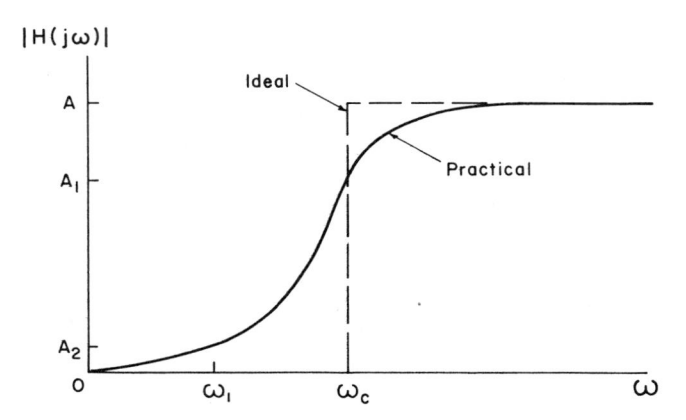

FIGURE 4-1 *Ideal and practical high-pass responses.*

where ω_c is the cutoff frequency and B and C are the normalized second-order stage low-pass coefficients given in Appendix A. In the odd-order case there is also a first-order section having transfer function

$$H = \frac{V_2}{V_1} = \frac{Ks}{s + \omega_c/C} \qquad (4\text{-}2)$$

where C is the normalized first-order low-pass coefficient.

A Butterworth high-pass filter has a monotonic response like the realizable case of Fig. 4-1, whereas the Chebyshev high-pass filter response is characterized by passband ripples. For example, a 1-dB Chebyshev high-pass filter, like its low-pass counterpart, has a 1-dB passband ripple channel. As an illustration, the response of an actual seventh-order 0.5-dB Chebyshev high-pass filter is shown in Fig. 4-2.

The *gain* of a high-pass filter is the value of its transfer function as s becomes infinite. Thus in the case of the second-order stage of Eq. (4-1) or the first-order stage of Eq. (4-2), the stage gain is K.

Inverse Chebyshev and elliptic high-pass filters have second-order sections with transfer functions of the form

$$\frac{V_2}{V_1} = \frac{K(s^2 + \omega_c^2/A)}{s^2 + (B\omega_c/C)s + \omega_c^2/C} \qquad (4\text{-}3)$$

where A, B, and C are the normalized low-pass coefficients of Appendix B or C. Odd-order filter functions have a first-order

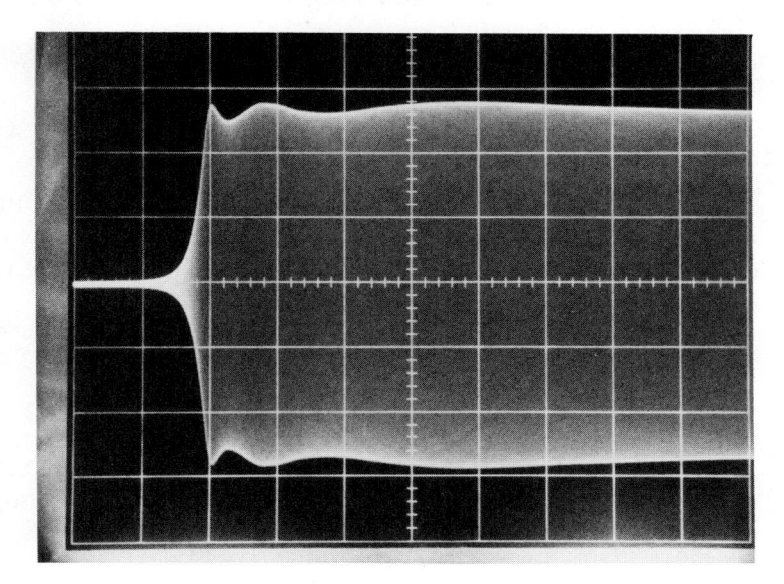

FIGURE 4-2 Chebyshev high-pass response.

stage with transfer function

$$\frac{V_2}{V_1} = \frac{Ks}{s + \omega_c/C} \tag{4-4}$$

where C is the low-pass coefficient of the first-order stage of Appendix B or C. In both Eqs. (4-3) and (4-4) K is the stage gain.

In both the high-pass second-order Butterworth or Chebyshev case Eq. (4-1) or the elliptic and inverse Chebyshev case Eq. (4-3), the pole-pair Q is defined, as in the low-pass filter, by $Q = \sqrt{C/B}$.

The inverse Chebyshev high-pass amplitude is monotonic in its passband and has stopband ripples characterized by a minimum stopband loss (MSL) in dB. We shall take ω_c as the 3-dB point in every case. The elliptic high-pass amplitude has passband and stopband ripples, characterized, respectively, by a passband ripple width (PRW) and a minimum stopband loss (MSL) in dB.

The transition width TW$_{\text{HP}}$ of the high-pass filter is related to the normalized transition width TW of its low-pass counterpart by

$$\text{TW}_{\text{HP}} = \frac{\text{TW}}{1 + \text{TW}} \omega_c \quad \text{rad/s} \tag{4-5}$$

or

$$\mathrm{TW_{HP}} = \frac{\mathrm{TW}}{1 + \mathrm{TW}} f_c \quad \mathrm{Hz} \qquad (4\text{-}6)$$

For example, a sixth-order elliptic high-pass filter with PRW = 0.5 dB and MSL = 60 dB has for its normalized low-pass counterpart a value of TW = 0.4014 (from Appendix D). Thus if $f_c = 1000$ Hz, its transition width, by Eq. (4-6), is

$$\mathrm{TW_{HP}} = \frac{0.4014}{1.4014}(1000)$$

$$= 286 \text{ Hz}$$

Alternatively, we may find TW for a given $\mathrm{TW_{HP}}$ from Eq. (4-5) or (4-6), resulting in

$$\mathrm{TW} = \frac{1}{(\omega_c/\mathrm{TW_{HP}}) - 1} \qquad (4\text{-}7)$$

The minimum order of a high-pass filter whose transition width does not exceed $\mathrm{TW_{HP}}$ is then the same as that of a normalized low-pass filter whose transition width does not exceed TW. For example, if we want the minimum order of a high-pass elliptic filter with $f_c = 1000$ Hz, PRW = 0.5 dB, MSL = 60 dB, and $\mathrm{TW_{HP}}$ not to exceed 100 Hz, then, by Eq. (4-7), we have

$$\mathrm{TW} = \frac{1}{1000/100 - 1} = 0.1111$$

Thus for the normalized low-pass case the transition width is not to exceed 0.1111. By Appendix D we find the minimum order to be $n = 9$.

4-2 INFINITE-GAIN MULTIPLE-FEEDBACK HIGH-PASS FILTERS

Like its low-pass counterpart, the second-order high-pass Butterworth or Chebyshev filter may be realized by an infinite-gain multiple-feedback (MFB) circuit, a VCVS circuit, and a biquad

SEC. 4-2   INFINITE-GAIN MFB HIGH-PASS FILTERS

circuit. In this section we shall consider the MFB circuit [9] and consider the other two types in the next two sections.

The MFB filter of Fig. 4-3 realizes the second-order high-pass

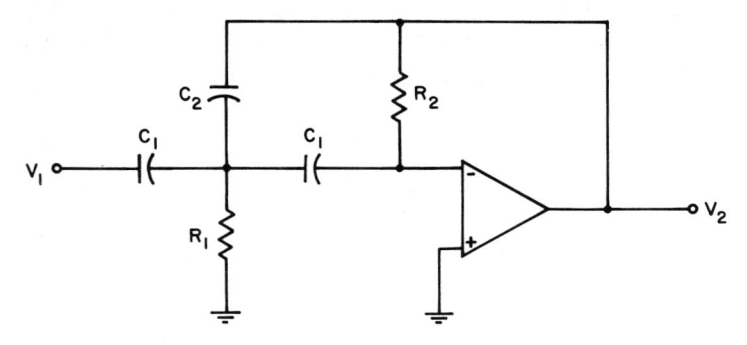

FIGURE 4-3 Infinite-gain MFB high-pass filter.

function (4-1) with inverting gain $-K$ ($K > 0$) for

$$K = \frac{C_1}{C_2}$$
$$\frac{B\omega_c}{C} = \frac{2C_1 + C_2}{R_2 C_1 C_2} \qquad (4\text{-}8)$$
$$\frac{\omega_c^2}{C} = \frac{1}{R_1 R_2 C_1 C_2}$$

A solution for the element values is given by

$$C_2 = \frac{C_1}{K}$$
$$R_1 = \frac{B}{(2C_1 + C_2)\omega_c} \qquad (4\text{-}9)$$
$$R_2 = \frac{(2C_1 + C_2)C}{B C_1 C_2 \omega_c}$$

where C_1 is arbitrary. Thus we may select C_1 (preferably near $10/f_c$ μF) and determine C_2 and the resistances. If $1/K$ is a standard capacitance multiple such as 1, 2, or $\frac{1}{2}$, C_2 will also be a standard value, such as C_1, $2C_1$, or $C_1/2$.

The relative merits of the MFB high-pass filter are the same as those of its low-pass counterpart listed in Section 2-5. The design procedure is summarized in Section 4-8.

4-3 VCVS HIGH-PASS FILTERS

A VCVS circuit that realizes the second-order high-pass Butterworth or Chebyshev filter function (4-1) is shown in Fig. 4-4 [26]. An analysis of the circuit yields

$$K = 1 + \frac{R_4}{R_3}$$

$$\frac{B\omega_c}{C} = \frac{1}{R_1 C_1}(1 - K) + \frac{2}{R_2 C_1} \quad (4\text{-}10)$$

$$\frac{\omega_c^2}{C} = \frac{1}{R_1 R_2 C_1^2}$$

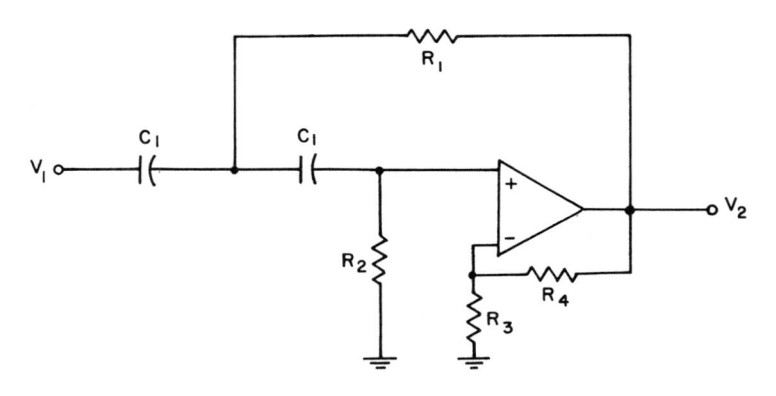

FIGURE 4-4 VCVS high-pass filter.

The gain is noninverting and the resistance values are given by

$$R_2 = \frac{4C}{[B + \sqrt{B^2 + 8C(K-1)}]\omega_c C_1}$$

$$R_1 = \frac{C}{\omega_c^2 C_1^2 R_2}$$

$$R_3 = \frac{KR_2}{K-1} \quad (4\text{-}11)$$

$$R_4 = KR_2$$

where C_1 is arbitrary.

SEC. 4-4   BIQUAD HIGH-PASS FILTERS   **75**

If $K = 1$, we may take R_3 as an open circuit and R_4 as a short circuit, in which case the op amp is functioning as a voltage follower. The resistances R_1 and R_2 remain the same.

The advantages of the VCVS high-pass circuit are the same as those of the VCVS low-pass circuit given in Section 2-6. The design procedure is summarized in Section 4-9.

4-4 BIQUAD HIGH-PASS FILTERS

A second-order biquad circuit that realizes the Butterworth or Chebyshev high-pass filter with inverting gain is shown in Fig. 4-5 [8]. Analysis of the circuit yields

$$K = \frac{R_5}{R_4}$$

$$\frac{B\omega_c}{C} = \frac{1}{R_2 C_1} \qquad (4\text{-}12)$$

$$\frac{\omega_c^2}{C} = \frac{1}{R_3 R_5 C_1^2}$$

where

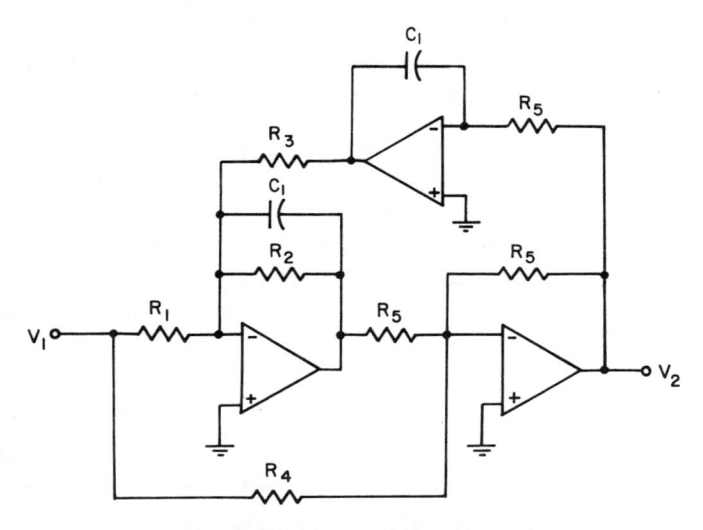

$$R_1 R_5 = R_2 R_4 \qquad (4\text{-}13)$$

FIGURE 4-5 Biquad high-pass filter.

The resistance values are given by

$$R_1 = \frac{C}{BK\omega_c C_1}$$
$$R_2 = KR_1$$
$$R_3 = \frac{C}{C_1^2 \omega_c^2 R_5} \quad (4\text{-}14)$$
$$R_4 = \frac{R_5}{K}$$

where C_1 and R_5 are arbitrary.

As in the low-pass case, the high-pass biquad has more elements than the MFB or VCVS filters. However, this disadvantage is offset by the excellent tuning features and stability of the biquad circuit.

The advantages of the high-pass biquad are the same as those of the low-pass biquad given in Section 2-7. The general design procedure is summarized in Section 4-10.

4-5 INVERSE CHEBYSHEV AND ELLIPTIC HIGH-PASS FILTER CIRCUITS

A second-order stage of an inverse Chebyshev or elliptic high-pass filter has transfer function given by

$$\frac{V_2}{V_1} = \frac{K(s^2 + \omega_c^2/A)}{s^2 + (B\omega_c/C)s + \omega_c^2/C} \quad (4\text{-}15)$$

where K is the stage gain and A, B, and C are the normalized coefficients of the low-pass counterpart of Appendix B or C. This function is identical to the low-pass transfer function (3-16) with

$$\rho = K, \quad \alpha = \frac{1}{A}, \quad \beta = \frac{B}{C}, \quad \gamma = \frac{1}{C} \quad (4\text{-}16)$$

Therefore, we may make these changes and use the results and circuits of Sections 3-3, 3-4, and 3-5 to realize the high-pass stage. This will be done in the design summaries given in Sections 4-11, 4-12, and 4-13.

4-6 TUNING THE SECOND-ORDER FILTERS

The second-order Butterworth or Chebyshev high-pass stage has an amplitude response which exhibits peaking, as shown in Fig. 4-6(a) if $Q = \sqrt{C}/B > \sqrt{2} = 1.414$. If $Q \leq 1.414$, there is no

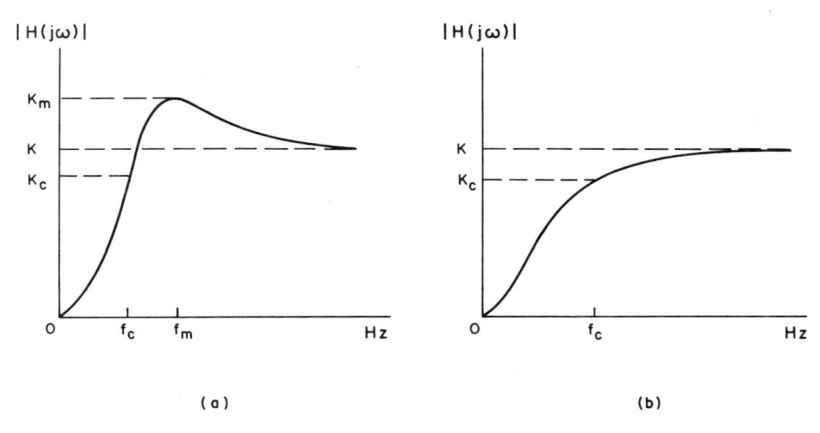

FIGURE 4-6 High-pass amplitude responses for: (a) $Q > 1.414$ and (b) $Q \leq 1.414$.

peaking and the response is as shown in Fig. 4-6(b). In Fig. 4-6(a) the peak K_m and the frequency f_m (Hz) at which it occurs are given by

$$K_m = \frac{2KC}{B\sqrt{4C - B^2}} \qquad (4\text{-}17)$$

$$f_m = \frac{f_c}{\sqrt{C - B^2/2}} \qquad (4\text{-}18)$$

In both cases the amplitude K_c at f_c is given by

$$K_c = \frac{KC}{\sqrt{(C-1)^2 + B^2}} \qquad (4\text{-}19)$$

Tuning is accomplished by adjusting the circuit element values until the amplitude response resembles Fig. 4-6(a) or (b). Suggestions are given in the design procedures for the various filter types in Sections 4-8 through 4-13.

In the case of the second-order inverse Chebyshev or elliptic high-pass stages, the amplitude response resembles Fig. 4-7(a) in the case of peaking. If there is no peaking, the response is like that of Fig. 4-7(b). In the case of peaking, the peak value K_m and the frequency f_m (Hz) at which it occurs are given by

$$K_m = \frac{2KC}{AB}\sqrt{\frac{(A-C)^2 + AB^2}{4C - B^2}} \qquad (4\text{-}20)$$

and

$$f_m = f_c\sqrt{\frac{2(A-C) + B^2}{2C(A-C) - AB^2}} \qquad (4\text{-}21)$$

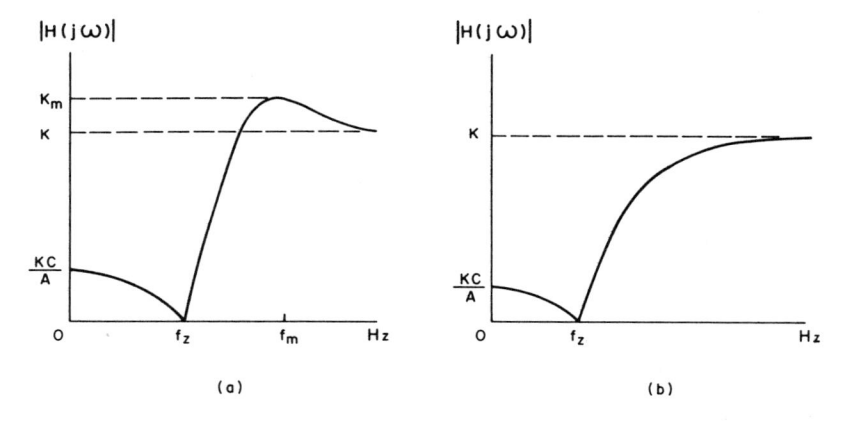

FIGURE 4-7 High-pass inverse Chebyshev or elliptic responses with: (a) peaking and (b) without peaking.

There is no peaking, of course, if f_m is imaginary.

In both figures the notch frequency is

$$f_z = \frac{f_c}{\sqrt{A}} \qquad (4\text{-}22)$$

and the value of the amplitude at the cutoff frequency f_c is given by

$$K_c = \frac{KC}{A}\frac{|A - 1|}{\sqrt{(C-1)^2 + B^2}} \qquad (4\text{-}23)$$

4-7 ODD-ORDER HIGH-PASS FILTERS

In the case of an odd-order high-pass filter of Butterworth, Chebyshev, inverse Chebyshev, or elliptic type, there will be one first-order stage with a transfer function of the form

$$\frac{V_2}{V_1} = \frac{Ks}{s + \omega_c/C} \qquad (4\text{-}24)$$

The coefficient C is that of the first-order low-pass stage given in Appendix A, B, or C, and K is the stage gain.

A circuit that achieves Eq. (4-24) for a gain $K > 1$ is shown in Fig. 4-8. The capacitance C_1 is arbitrary and the resistances are

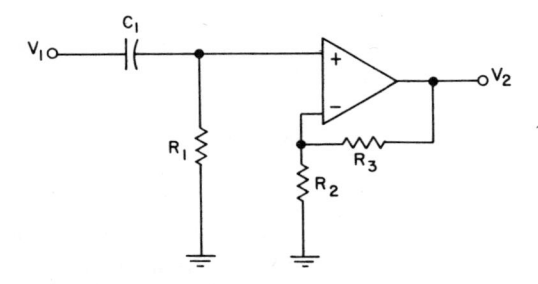

FIGURE 4-8 First-order high-pass filter circuit.

given by

$$R_1 = \frac{C}{\omega_c C_1}$$

$$R_2 = \frac{KR_1}{K-1} \qquad (4\text{-}25)$$

$$R_3 = KR_1$$

If a gain of $K = 1$ is desired, we may take R_1 as in Eq. (4-25) and replace R_2 by an open circuit and R_3 by a short circuit. In this case the circuit is a voltage-follower circuit.

We may realize the remaining stages, which are all second order, by the methods of Section 4-2, 4-3, 4-4, or 4-5, and cascade the stages to form the filter.

4-8 MFB HIGH-PASS FILTER DESIGN SUMMARY

To design a second-order high-pass filter, or a second-order stage of a higher-order filter, having a given cutoff frequency f_c Hz (or $\omega_c = 2\pi f_c$ rad/s), gain K, and of Butterworth or Chebyshev type, perform the following steps.

1. Find the normalized low-pass coefficients B and C from the appropriate table of Appendix A
2. Select a standard value of C_1 (preferably near $10/f_c \, \mu F$) and calculate the element values given by

$$C_2 = \frac{C_1}{K}$$

$$R_1 = \frac{B}{(2C_1 + C_2)\omega_c}$$

$$R_2 = \frac{(2C_1 + C_2)C}{BC_1 C_2 \omega_c}$$

3. Select standard values of the elements as close as possible to the calculated values and construct the filter, or its stages, in accordance with Fig. 4-9. It should be noted that the gain K is restricted to a ratio of standard capacitances, or else trimming of the capacitance C_2 will be required

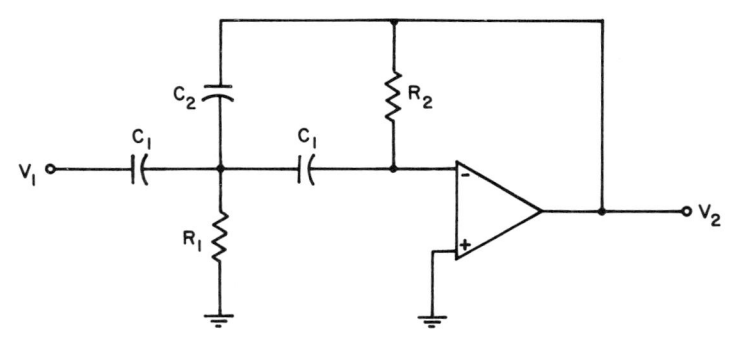

FIGURE 4-9 MFB high-pass filter circuit.

Comments

(a) For best performance, element values close to those selected and calculated should be used. Higher-order filters require more accurate element values than lower-order filters. The performance of the filter is unchanged if all the resistances are multiplied and the capacitances divided by a common factor

(b) The input impedance of the op amp should be at least $10R_{eq}$, where

$$R_{eq} = R_2$$

The open-loop gain of the op amp should be at least 50 times the amplitude of the filter, or stage, at f_a, the highest desired frequency in the passband, and its slew rate (volts per microsecond) should be at least $\frac{1}{2}\omega_a \times 10^{-6}$ times the peak-to-peak output voltage

(c) Resistor R_1 or R_2 may be adjusted to set the peak at f_m (see Section 4-6). Afterward, R_1 and R_2 may be adjusted simultaneously by the same percentage, without changing the pole-pair $Q = \sqrt{C/B}$, to set the cutoff at f_c. These steps may be repeated if necessary

(d) The gain is inverting with magnitude $K = C_1/C_2$. Thus gain adjustments may be made by trimming either C_1 or C_2

(e) The circuit should be used only for filter stages with gain K and pole-pair Q of 10 or less. The gain could be higher if Q is lower, with limits of $KQ = 100$ and $Q = 10$

(f) The order required for a given transition width, or conversely, the transition width ensuing from a given order, may be found by Section 4-1

The MFB high-pass filter was discussed in Section 4-2.

4-9 VCVS HIGH-PASS FILTER DESIGN SUMMARY

To design a second-order high-pass filter, or a second-order stage of a higher-order filter, having a given cutoff frequency f_c Hz (or $\omega_c = 2\pi f_c$ rad/s), gain $K \geq 1$, and of Butterworth or Chebyshev type, perform the following steps.

1. Find the normalized low-pass coefficients B and C from the appropriate table of Appendix A

2. Select a standard value of C_1 (preferably near $10/f_c$ μF) and calculate the resistance values given by

$$R_2 = \frac{4C}{[B + \sqrt{B^2 + 8C(K-1)}]\omega_c C_1}$$

$$R_1 = \frac{C}{\omega_c^2 C_1^2 R_2}$$

$$R_3 = \frac{KR_2}{K-1}$$

$$R_4 = KR_2$$

3. Select standard values of the resistances as close as possible to the calculated values and construct the filter, or its stages, in accordance with Fig. 4-10

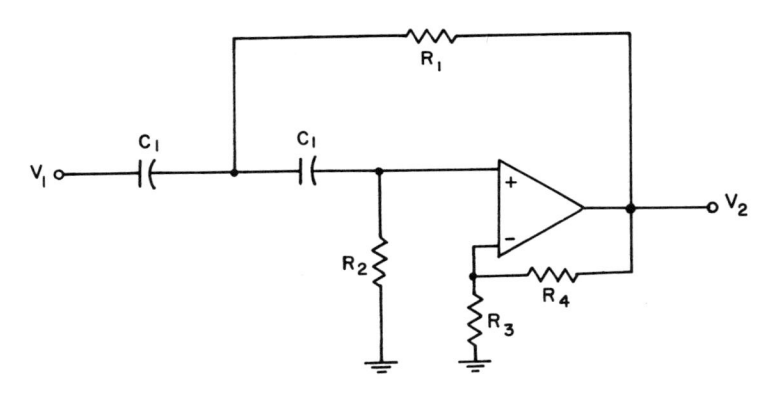

FIGURE 4-10 VCVS high-pass filter circuit.

Comments

(a) Comments (a), (b), (e), and (f) for the MFB filter of Section 4-8 apply directly

(b) The values of R_3 and R_4 are for $K > 1$ and are chosen to minimize the dc offset of the op amp. The section gain is a noninverting one given by

$$K = 1 + \frac{R_4}{R_3}$$

so other values of R_3 and R_4 may be used as long as their ratio is maintained. If $K = 1$ is desired, R_3 becomes an open circuit and R_4 becomes a short circuit, in which case the circuit is of the voltage-follower type

(c) Resistors R_1 and R_2 may be adjusted by equal percentages, without affecting the pole-pair Q, to set the cutoff at f_c (see Section 4-6). The gain K may be set by using a potentiometer in lieu of resistors R_3 and R_4 with the center tap connected to the inverting input of the op amp

The VCVS high-pass filter was discussed in Section 4-3.

4-10 BIQUAD HIGH-PASS FILTER DESIGN SUMMARY

To design a second-order high-pass filter, or a second-order stage of a higher-order filter, having a given cutoff frequency f_c Hz (or $\omega_c = 2\pi f_c$ rad/s), gain K, and of Butterworth or Chebyshev type, perform the following steps.

1. Find the normalized low-pass coefficients B and C from the appropriate table of Appendix A

2. Select a standard value of C_1 (preferably near $10/f_c$ μF), and calculate the resistance values given by

$$R_1 = \frac{C}{BK\omega_c C_1}$$
$$R_2 = KR_1$$
$$R_3 = \frac{C}{C_1^2 \omega_c^2 R_5}$$
$$R_4 = \frac{R_5}{K}$$

where R_5 is arbitrary, with a reasonable value of

$$R_5 = \frac{1}{\omega_c C_1}$$

3. Select standard values of resistance as close as possible to the calculated values and construct the filter, or its stages, in accordance with Fig. 4-11.

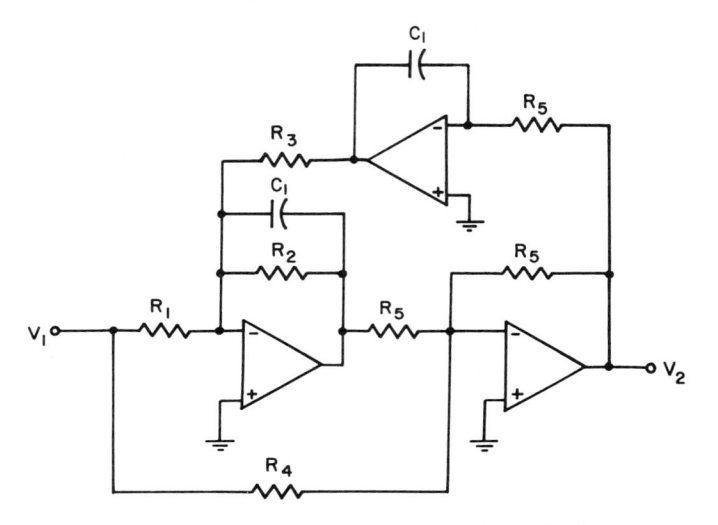

FIGURE 4-11 Biquad high-pass filter circuit.

Comments

(a) Comments (a), (b), and (f) for the MFB filter of Section 4-8 apply directly, except that in (b), R_{eq} for each op amp is the resistance R_1 or R_5 connected to its inverting input terminal

(b) The stage gain is inverting with magnitude $K = R_2/R_1$

(c) Tuning may be accomplished by varying R_1 or R_4 to adjust K, varying R_3 to adjust f_c, and varying R_2 to adjust the passband response (see Section 4-6). These steps may be repeated if necessary. The value of R_5 may be chosen arbitrarily to maintain a smaller spread of resistance values

(d) The filter may be used for pole-pair Q values up to 100

The biquad high-pass filter was discussed in Section 4-4.

4-11 VCVS HIGH-PASS ELLIPTIC FILTER DESIGN SUMMARY

To design a second-order high-pass elliptic or inverse Chebyshev filter, or a second-order stage of a higher-order filter, having a given cutoff frequency f_c Hz (or $\omega_c = 2\pi f_c$ rad/s), gain K, minimum stopband loss (MSL), and in the case of the elliptic filter, passband ripple width (PRW), perform the following steps.

1. Find the normalized low-pass coefficients A, B, and C from the appropriate table of Appendix B or C

2. Select a standard value of C_1 (preferably near $10/f_c$ μF) and calculate the resistance values given by

$$R_1 = \frac{\mu AB}{KC\omega_c C_1}$$

$$R_2 = \frac{C}{B\omega_c C_2}$$

$$R_3 = \frac{KCR_1}{A}$$

$$R_4 = \frac{KR_5}{\mu}$$

$$R_6 = \frac{\mu R_2}{\mu - 1}$$

$$R_7 = \mu R_2$$

where C_2, R_5, and $\mu > 1$ are arbitrary. If K and the pole-pair $Q = \sqrt{C}/B$ are moderate values, such as 10 or less, reasonable values of the arbitrary parameters are

$$C_2 = C_1$$

$$R_5 = \frac{1}{\omega_c C_1}$$

and $\mu = 2$ (in which case $R_6 = R_7$). If Q and/or K are high, say over 10, C_2, R_5, and μ should be chosen to maintain a lower spread of resistance values

3. Select standard values of resistances as close as possible to the calculated values and construct the filter, or its stages, in accordance with Fig. 4-12

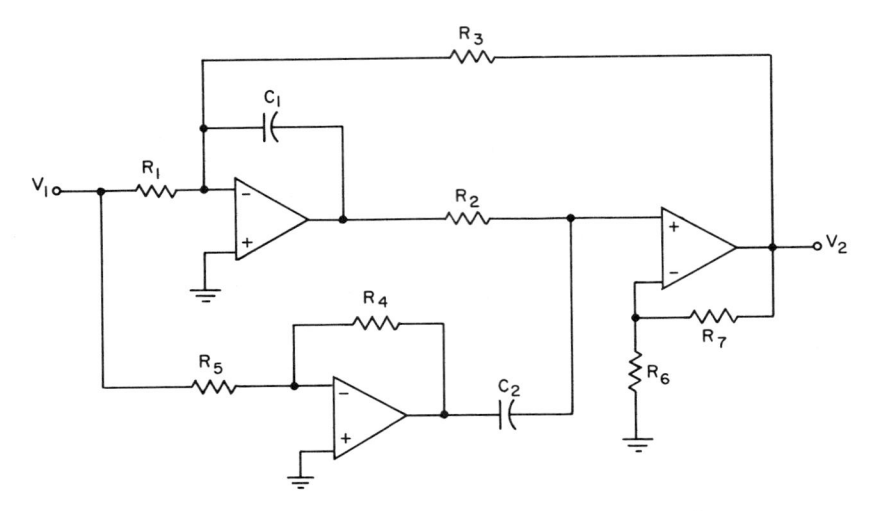

FIGURE 4-12 *VCVS high-pass elliptic filter circuit.*

4. If $\mu = 1$ is desired, R_6 becomes an open circuit and R_7 becomes a short circuit, with the other resistances given as in step 2. In this case the circuit is the voltage follower circuit of Fig. 3-6

Comments

(a) Comments (a), (b), and (f) for the MFB filter of Section 4-8 apply directly, except that in (b), R_{eq} for each op amp is the resistance R_1, R_2, or R_5 connected to its inverting input terminal

(b) The circuit may be used for both high and low pole-pair Q by selecting the arbitrary parameters of step 2 to keep the spread of resistance values relatively low. For moderate gains, Q may range as high as 100

(c) Tuning may be accomplished by varying the ratio R_4/R_5 to move the notch toward f_z, shown in Fig. 4-7. The gain μ of the VCVS, given by

$$\mu = 1 + \frac{R_7}{R_6}$$

then may be adjusted, by varying the ratio R_7/R_6, to move the peak toward f_m. Finally, R_2 may be varied to obtain K_m. These steps may be repeated until the stage is tuned

SEC. 4-12   3-CAPACITOR HP ELLIPTIC FILTER DESIGN   **87**

(d) The circuit yields an inverting gain with magnitude $K = \mu R_4/R_5$

The VCVS high-pass elliptic filter was discussed in Section 4-5.

4-12 THREE-CAPACITOR HIGH-PASS ELLIPTIC FILTER DESIGN SUMMARY

To design a second-order high-pass elliptic or inverse Chebyshev filter, or a second-order stage of a higher-order filter, having a given cutoff frequency f_c Hz (or $\omega_c = 2\pi f_c$ rad/s), gain K, minimum stopband loss (MSL), and in the case of the elliptic filter, passband ripple width (PRW), perform the following steps.

1. Find the normalized low-pass coefficients A, B, and C from the appropriate table of Appendix B or C

2. Select a standard value of C_1 (preferably near $10/f_c$ μF) and calculate the element values given by

$$C_3 = KC_2$$

$$R_1 = \frac{A}{R_4 \omega_c^2 C_1 C_3}$$

$$R_2 = \frac{KCR_1}{A}$$

$$R_3 = \frac{C}{B\omega_c C_2}$$

where C_2 and R_4 are arbitrary. If the pole-pair $Q = \sqrt{C/B}$ is low, C_2 may be chosen near C_1, and if Q is high, C_2 may be chosen larger than C_1. In either case, R_4 may be selected to minimize the spread of the resistance values

3. Select standard values of resistances and capacitances as close as possible to the calculated values and construct the filter, or its stages, in accordance with Fig. 4-13

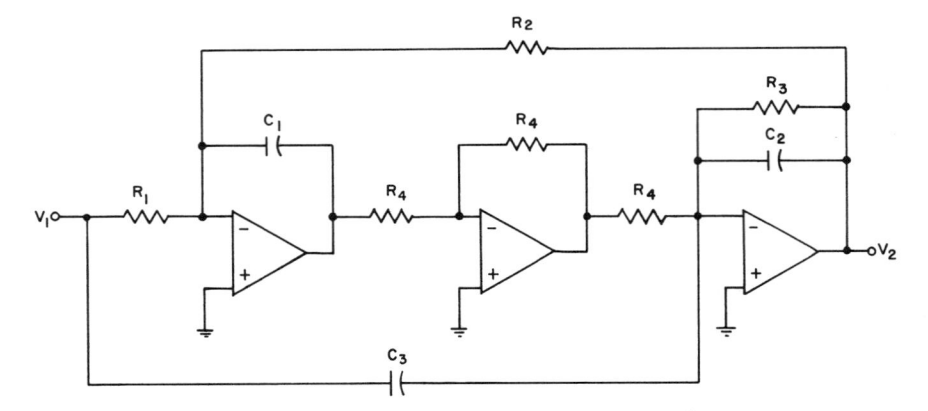

FIGURE 4-13 Three-capacitor high-pass elliptic filter circuit.

Comments

(a) Comments (a), (b), and (f) for the MFB filter of Section 4-8 apply directly, except that in (b), R_{eq} for each op amp is the resistance R_1 or R_4 connected to its inverting input

(b) The circuit may be used for both low and high Q, with an upper limit of approximately $Q = 100$

(c) Tuning may be accomplished by first adjusting R_1 until the notch is at f_z (see Fig. 4-7), then adjusting R_2 to place the peak amplitude at f_m, and finally adjusting R_3 to obtain the correct value of K_m. For low-Q sections, the last two steps may need to be repeated

(d) The circuit yields an inverting gain with magnitude $K = C_3/C_2$

The three-capacitor circuit was discussed in Section 4-5.

4-13 BIQUAD HIGH-PASS ELLIPTIC FILTER DESIGN SUMMARY

To design a second-order high-pass elliptic or inverse Chebyshev filter, or a second-order stage of a higher-order filter, having a given cutoff frequency f_c Hz (or $\omega_c = 2\pi f_c$ rad/s), gain K, minimum stopband loss (MSL), and in the case of the elliptic filter, passband ripple width (PRW), perform the following steps.

SEC. 4-13 BIQUAD HIGH-PASS ELLIPTIC FILTER DESIGN

1. Find the normalized low-pass coefficients A, B, and C from the appropriate table of Appendix B or C
2. Select a standard value of C_1 (preferably near $10/f_c$ μF) and calculate the element values given by

$$R_1 = \frac{C}{KB\omega_c C_1}$$

$$R_2 = KR_1$$

$$R_3 = \frac{\sqrt{C}}{\omega_c C_1}$$

$$R_4 = \frac{R_7}{K}$$

$$R_5 = \frac{A}{K\sqrt{C}\omega_c C_2}$$

$$R_6 = \frac{C_1 R_3}{C_2}$$

The values of C_2 and R_7 are arbitrary and may be chosen, depending on the gain K and the pole-pair $Q = \sqrt{C}/B$, to minimize the spread of resistance values. For moderate K and Q, reasonable values are $C_2 = C_1$ and $R_7 = 1/\omega_c C_1$.

3. Select standard values of resistances and capacitances as close as possible to the calculated values and construct the filter, or its stages, in accordance with Fig. 4-14

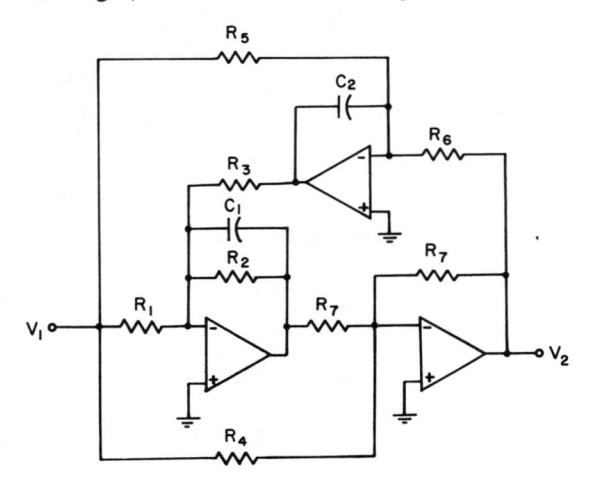

FIGURE 4-14 Biquad high-pass elliptic filter circuit.

Comments

(a) Comments (a), (b), and (f) for the MFB filter of Section 4-8 apply directly, except that in (b), R_{eq} for each op amp is the resistance R_1, R_6, or R_7 connected to its inverting input

(b) The circuit may be used for both low and high pole-pair Q, with an upper limit of approximately $Q = 100$

(c) Tuning may be accomplished by adjusting R_4 to set the notch at f_z, adjusting R_3 to set the cutoff frequency, adjusting R_2 to set Q, and adjusting R_1 to set the gain

(d) The circuit yields an inverting gain with magnitude R_2/R_1

The biquad elliptic filter was discussed in Section 4-5.

4-14 ODD-ORDER HIGH-PASS FILTER DESIGN SUMMARY

To design a first-order stage of a higher odd-order high-pass filter of a given type (Butterworth, Chebyshev, inverse Chebyshev, or elliptic) having a given cutoff frequency f_c Hz (or $\omega_c = 2\pi f_c$ rad/s), stage gain K, minimum stopband loss (MSL), and passband ripple width (PRW), if applicable, perform the following steps.

1. Find the normalized low-pass coefficient C for the first-order stage from the appropriate table of Appendix A, B, or C

2. Select a standard value of capacitance C_1 (preferably near $10/f_c$ μF)

3. (a) If $K > 1$, calculate the resistance values given by

$$R_1 = \frac{C}{\omega_c C_1}$$

$$R_2 = \frac{KR_1}{K-1}$$

$$R_3 = KR_1$$

and construct the filter in accordance with Fig. 4-15

SEC. 4-14 ODD-ORDER HIGH-PASS FILTER DESIGN

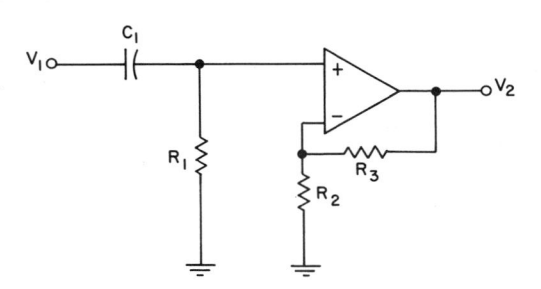

FIGURE 4-15 First-order high-pass filter circuit.

(b) If $K = 1$, use R_1 in step 3(a) and replace R_2 by an open circuit and R_3 by a short circuit. This results in a voltage-follower circuit

4. The second-order stages of the odd-order filter may be constructed by the methods of the previous sections and cascaded with the first-order stage to form the filter

Comments

(a) Comments (a) and (b) for the MFB filter of Section 4-8 apply directly, except that in (b), $R_{eq} = R_1$

(b) The values of R_2 and R_3 are chosen to minimize the dc offset of the op amp. Other values may be used as long as their ratio is maintained, in which case

$$K = 1 + \frac{R_3}{R_2}$$

(c) The gain $K > 1$ may be adjusted by using a potentiometer in lieu of resistors R_2 and R_3 with the center tap connected to the inverting input of the op amp. The cutoff frequency may be adjusted by varying R_1. The response should look like that of Fig. 4-6(b)

The first-order high-pass filter was discussed in Section 4-7.

5

BANDPASS FILTERS

5-1 THE GENERAL CASE

A *bandpass filter* is one that passes a band of frequencies of *bandwidth* BW centered approximately about a *center frequency* ω_0 rad/s, or $f_0 = \omega_0/2\pi$ Hz. An ideal and a practical bandpass amplitude response are shown in Fig. 5-1. In the practical case, shown with a solid line, the frequencies ω_L and ω_U are the *lower* and *upper cutoff frequencies*, which define the passband $\omega_L \leq \omega \leq \omega_U$ and the bandwidth BW $= \omega_U - \omega_L$.

In the passband the amplitude is never less than some prescribed value, such as A_1 in Fig. 5-1. Also, there are two stopbands, $0 \leq \omega \leq \omega_1$ and $\omega \geq \omega_2$, where the amplitude never exceeds a prescribed value, such as A_2. The regions between the stopbands and the passband, namely $\omega_1 < \omega < \omega_L$ and $\omega_U < \omega < \omega_2$, are, respectively, the *lower* and *upper transition bands* in which the response is monotonic.

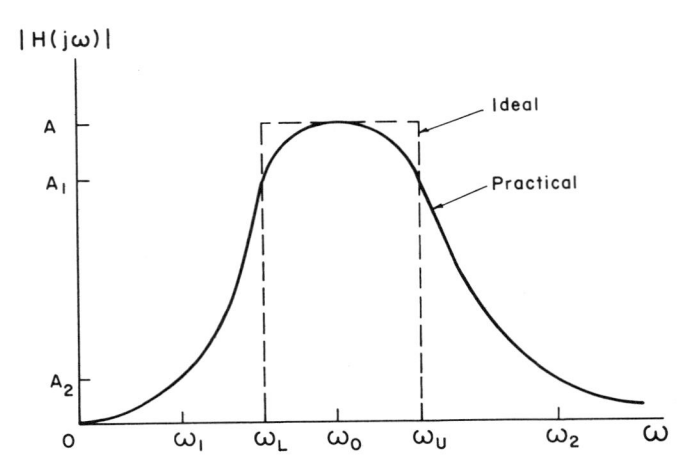

FIGURE 5-1 *Ideal and practical bandpass responses.*

The ratio $Q = \omega_0/\text{BW}$ is the *quality factor* of the filter and is a measure of its selectivity. High Q corresponds to a relatively narrow bandwidth and low Q corresponds to a relatively wide bandwidth. The *gain K* of the filter is the value of its amplitude at the center frequency; that is, $K = |H(j\omega_0)|$.

Bandpass transfer functions may be obtained from normalized low-pass functions of S by the transformation [16]

$$S = \frac{s^2 + \omega_0^2}{\text{BW} \cdot s} = \frac{Q(s^2 + \omega_0^2)}{\omega_0 s} \tag{5-1}$$

Consequently, the order of a bandpass filter is twice that of its corresponding low-pass filter, and thus is always even. The resulting bandpass amplitude has center frequency ω_0 and bandwidth BW, and resembles its low-pass counterpart shifted upward in frequency from 0 to ω_0. Thus a Butterworth bandpass amplitude (obtained from a low-pass Butterworth function) varies monotonically on either side of its peak, with a maximally flat passband, as in Fig. 5-1. A Chebyshev bandpass filter has passband ripples, an inverse Chebyshev bandpass filter has stopband ripples, and an elliptic bandpass filter has both pass- and stopband ripples. In every case the center and cutoff frequencies are related by

$$\omega_0 = \sqrt{\omega_L \omega_U}$$

SEC. 5-1 THE GENERAL CASE

where

$$\omega_L = \omega_0 \left(-\frac{1}{2Q} + \sqrt{1 + \frac{1}{4Q^2}}\right)$$
$$\omega_U = \omega_0 \left(\frac{1}{2Q} + \sqrt{1 + \frac{1}{4Q^2}}\right)$$
(5-2)

Examples of fourth-order Butterworth and fourth-order 1-dB Chebyshev bandpass amplitudes are shown, for $\omega_0 = 1$ rad/s and various values of Q, in Figs. 5-2 and 5-3. It is clear from these results that higher Q corresponds to narrower passbands.

Examples of bandpass responses of actual circuits are shown in Figs. 5-4 and 5-5. The responses are those of a fourth-order Butterworth and a fourth-order 0.5-dB Chebyshev filter, respectively, both with $Q = 5$.

A fourth-order response of an actual elliptic bandpass filter is shown in Fig. 5-6. It has a Q of 10, a passband ripple width of 0.5 dB, and a minimum stopband loss of 40 dB. Because the latter is so large, the stopband ripples are difficult to see, so that, except for the sharpness of cutoff, the response looks much like a Chebyshev response.

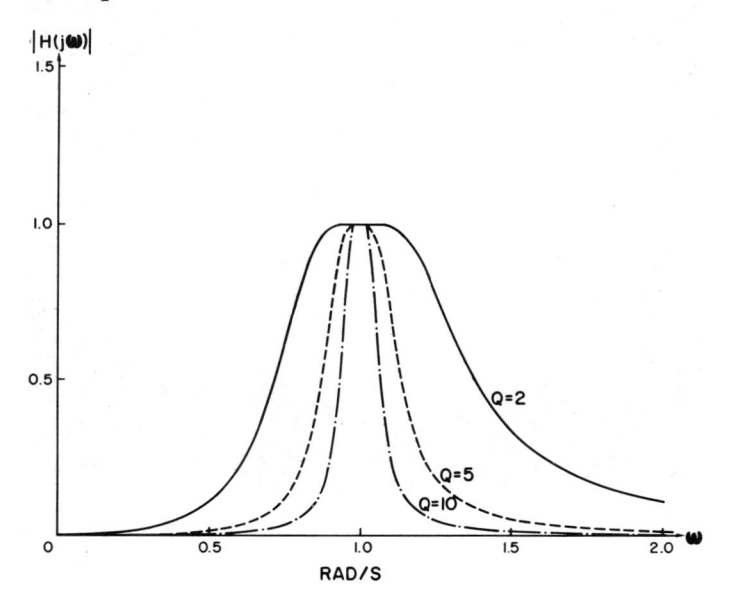

FIGURE 5-2 *Fourth-order Butterworth bandpass responses.*

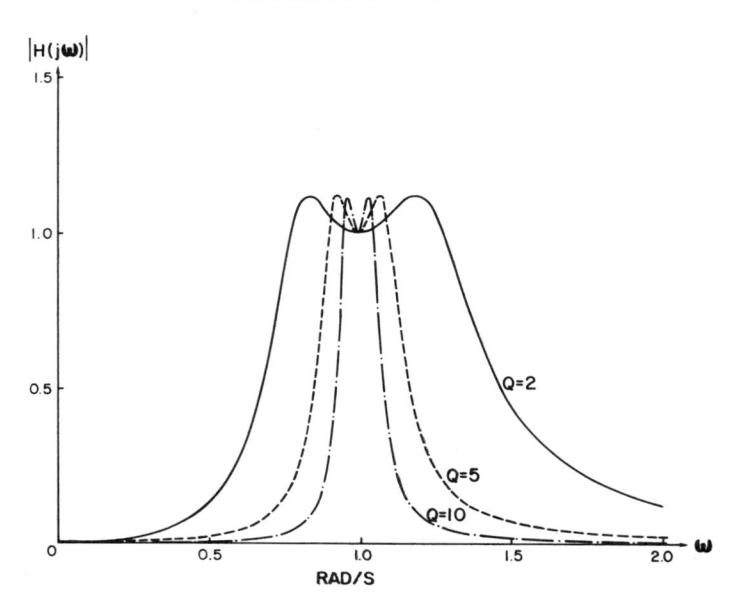

FIGURE 5-3 Fourth-order 1 dB Chebyshev bandpass responses.

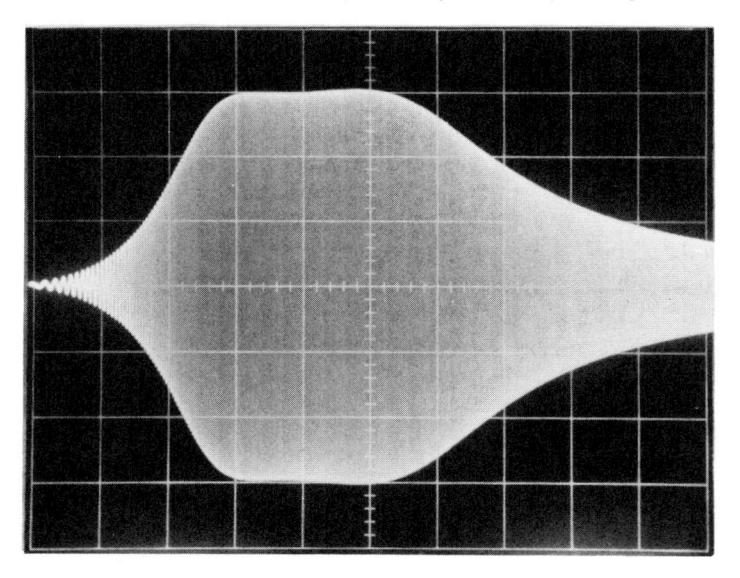

FIGURE 5-4 Fourth-order Butterworth response of an actual circuit.

SEC. 5-1　　　THE GENERAL CASE　　　97

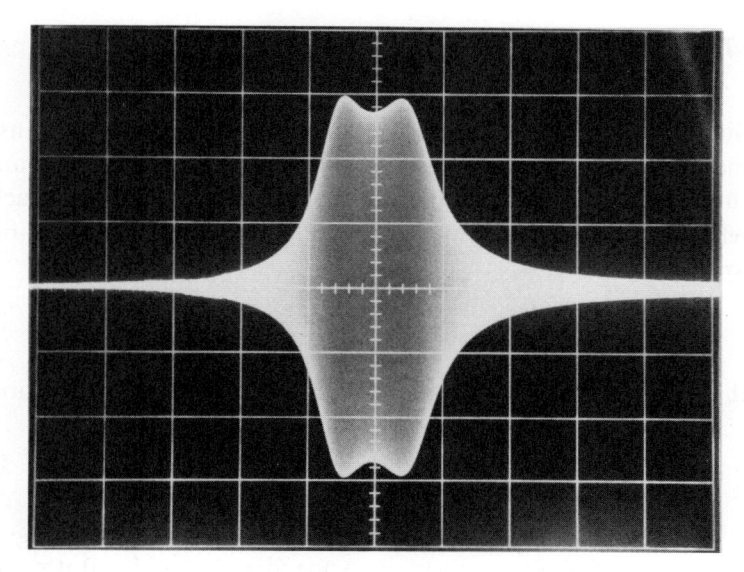

FIGURE 5-5 Fourth-order 0.5 dB Chebyshev response of an actual circuit.

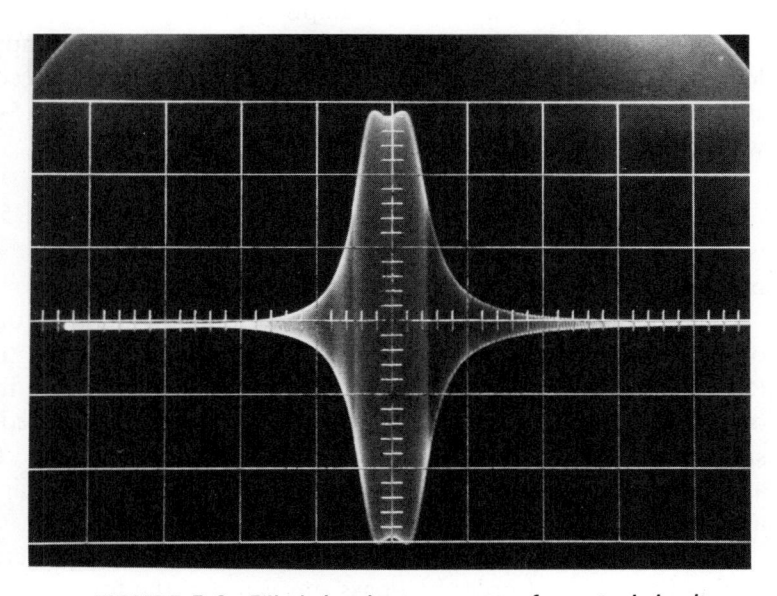

FIGURE 5-6 Elliptic bandpass response of an actual circuit.

5-2 TRANSFER FUNCTIONS

Since bandpass functions are obtained by applying the transformation (5-1) to the corresponding low-pass function, the transfer function of a bandpass filter will be a product of factors, each of which arises from a low-pass factor. In the case of a first-order low-pass factor,

$$\frac{V_2}{V_1} = \frac{KC}{S+C} \tag{5-3}$$

the corresponding bandpass factor is the second-order function

$$\frac{V_2}{V_1} = \frac{KC\omega_0 s/Q}{s^2 + (C\omega_0/Q)s + \omega_0^2} \tag{5-4}$$

where C is the normalized coefficient of the corresponding low-pass first-order stage, given in Appendix A for the Butterworth and Chebyshev filters, in Appendix B for the inverse Chebyshev filter, and in Appendix C for the elliptic filter.

A second-order bandpass filter results when the corresponding low-pass filter is of first order. That is, the low-pass function consists solely of Eq. (5-3) with $C = 1$. In this case, from Eq. (5-4), we have the transfer function

$$\frac{V_2}{V_1} = \frac{K\omega_0 s/Q}{s^2 + (\omega_0/Q)s + \omega_0^2} \tag{5-5}$$

of a second-order bandpass filter.

The transfer function (5-5) may be thought of as a second-order Butterworth or Chebyshev bandpass filter because Eq. (5-3), with $C = 1$, is a Butterworth or a scaled Chebyshev first-order low-pass function. Usually, however, Eq. (5-5) is referred to simply as a second-order bandpass filter transfer function, and the adjectives Butterworth, Chebyshev, inverse Chebyshev, and elliptic are reserved for higher-order bandpass filters.

Butterworth or Chebyshev bandpass transfer-function factors arising from second-order low-pass stages are of the form

$$\frac{V_2}{V_1} = \frac{(KC\omega_0^2/Q^2)s^2}{s^4 + (B\omega_0/Q)s^3 + (2 + C/Q^2)\omega_0^2 s^2 + (B\omega_0^3/Q)s + \omega_0^4} \tag{5-6}$$

SEC. 5-2 TRANSFER FUNCTIONS

where B and C are the corresponding low-pass coefficients of Appendix A. In Eq. (5-4) K is the stage gain, and in Eq. (5-6) K is the overall gain of two second-order stages cascaded to realize the fourth-order function.

The transfer function Eq. (5-6) may be factored into the two second-order functions [4]

$$\left(\frac{V_2}{V_1}\right)_1 = \frac{(K_1\omega_0\sqrt{C/Q})s}{s^2 + (D\omega_0/E)s + D^2\omega_0^2} \tag{5-7}$$

and

$$\left(\frac{V_2}{V_1}\right)_2 = \frac{(K_2\omega_0\sqrt{C/Q})s}{s^2 + (\omega_0/DE)s + \omega_0^2/D^2} \tag{5-8}$$

where

$$E = \frac{1}{B}\sqrt{\frac{C + 4Q^2 + \sqrt{(C + 4Q^2)^2 - (2BQ)^2}}{2}} \tag{5-9}$$

and

$$D = \frac{1}{2}\left[\frac{BE}{Q} + \sqrt{\left(\frac{BE}{Q}\right)^2 - 4}\right] \tag{5-10}$$

Thus the transfer function of a Butterworth or Chebyshev bandpass filter of order $n = 4, 6, 8, \ldots$ will have a factor like Eq. (5-7) and one like Eq. (5-8) for each second-order stage in its corresponding low-pass filter. The numbers K_1 and K_2 are the gains of the two bandpass stages and should be chosen so that $K_1 K_2 = K$.

In summary, a typical transfer function of a second-order bandpass filter or a second-order stage of a higher-order Butterworth or Chebyshev bandpass filter is of the form

$$\frac{V_2}{V_1} = \frac{\rho\omega_0 s}{s^2 + \beta\omega_0 s + \gamma\omega_0^2} \tag{5-11}$$

where ρ, β, and γ are obtained by matching Eq. (5-11) with the appropriate one of Eq. (5-4), (5-5), (5-7), or (5-8).

It is interesting to note that E is the pole-pair Q of each stage defined by Eqs. (5-7) and (5-8). As in the low-pass and high-pass cases, high Q usually requires more elaborate circuits.

To illustrate the use of Eq. (5-11), suppose that we want the transfer function of a fourth-order Butterworth bandpass filter with gain $K = 4$, center frequency $\omega_0 = 1$ rad/s, and $Q = 5$. We have from Appendix A, $B = 1.414214$ and $C = 1$. In this case, since the corresponding low-pass filter has only one section, of second order, there will be one factor for which Eq. (5-11) is Eq. (5-7) and one factor for which Eq. (5-11) is Eq. (5-8). Choosing arbitrarily $K_1 = K_2 = 2$, so that $K_1 K_2 = K$, the numerators of Eqs. (5-7) and (5-8) are determined. By Eq. (5-9) we have $E = 7.088812$, which, together with B and Q in Eq. (5-10), yields $D = 1.073397$. These values determine the denominators of Eqs. (5-7) and (5-8), which complete the determination of the transfer functions, given by

$$\left(\frac{V_2}{V_1}\right)_1 = \frac{0.4s}{s^2 + 0.151421s + 1.152181}$$

and

$$\left(\frac{V_2}{V_1}\right)_2 = \frac{0.4s}{s^2 + 0.131421s + 0.867919}$$

In the case of elliptic or inverse Chebyshev bandpass filters, the transfer function may also be factored into second-order functions. If the corresponding low-pass filter is of odd order, the bandpass factor ensuing from the first-order low-pass stage is Eq. (5-4), which is a special case of Eq. (5-11), as noted earlier. The two factors arising from each second-order low-pass stage have the forms

$$\left(\frac{V_2}{V_1}\right)_1 = \frac{K_1 \sqrt{C/A}(s^2 + A_1 \omega_0^2)}{s^2 + (D\omega_0/E)s + D^2 \omega_0^2} \tag{5-12}$$

and

$$\left(\frac{V_2}{V_1}\right)_2 = \frac{K_2 \sqrt{C/A}(s^2 + \omega_0^2/A_1)}{s^2 + (\omega_0/DE)s + \omega_0^2/D^2} \tag{5-13}$$

where E and D are given by Eqs. (5-9) and (5-10), and

$$A_1 = 1 + \frac{1}{2Q^2}(A + \sqrt{A^2 + 4AQ^2}) \tag{5-14}$$

The coefficients A, B, and C are those of the normalized low-pass functions of Appendix B or C, and K_1 and K_2 are the stage gains.

Equations (5-12) and (5-13) are of the general form

$$\frac{V_2}{V_1} = \frac{p(s^2 + \alpha\omega_0^2)}{s^2 + \beta\omega_0 s + \gamma\omega_0^2} \quad (5\text{-}15)$$

which is identical to the low-pass function (3-16) except that ω_0 has replaced ω_c.

5-3 TRANSITION WIDTHS

As was pointed out in Section 5-1, a bandpass filter has two stop-bands, $0 \leq \omega \leq \omega_1$ and $\omega \geq \omega_2$, where ω_1 and ω_2 are prescribed frequencies. There are also two transition bands, a lower transition band $\omega_1 < \omega < \omega_L$ of transition width

$$TW_L = \omega_L - \omega_1 \quad (5\text{-}16)$$

and an upper transition band $\omega_U < \omega < \omega_2$ of transition width

$$TW_U = \omega_2 - \omega_U \quad (5\text{-}17)$$

Applying the transformation (5-1) to the low-pass function to obtain the bandpass function also relates the transition widths of the bandpass filter to that of its corresponding low-pass filter. The results are

$$\frac{TW_L}{\omega_0} = \frac{1}{2Q}[TW - (\sqrt{(1 + TW)^2 + 4Q^2} - \sqrt{1 + 4Q^2})] \quad (5\text{-}18)$$

and

$$\frac{TW_U}{\omega_0} = \frac{1}{2Q}[TW + (\sqrt{(1 + TW)^2 + 4Q^2} - \sqrt{1 + 4Q^2})] \quad (5\text{-}19)$$

where TW is the normalized low-pass transition width of the corresponding low-pass filter. For high-Q circuits a good approximation to Eqs. (5-18) and (5-19) is

$$\frac{TW_L}{\omega_0} = \frac{TW_U}{\omega_0} = \frac{TW}{2Q} \quad (5\text{-}20)$$

which is the average of the two transition widths. The values of TW are given, in the case of Butterworth and Chebyshev filters, by Eqs. (2-16) and (2-17), with $\omega_c = 1$, and for inverse Chebyshev filters by Eq. (3-7). TW is tabulated for elliptic filters in Appendix C.

As an example, let us find the transition widths of an elliptic bandpass filter of order 8 with PRW $= 0.5$ dB, MSL $= 60$ dB, $f_0 = \omega_0/2\pi = 1000$ Hz, and $Q = 5$. From Appendix C we have TW $= 0.1243$ and from Eq. (5-18) we have

$$\frac{TW_L}{1000} = \frac{1}{10}[0.1243 - (\sqrt{1.1243^2 + 100} - \sqrt{1 + 100})]$$

or $TW_L = 11.1172$ Hz. Similarly, from Eq. (5-19) we have $TW_U = 13.7428$ Hz. The approximation of Eq. (5-20) yields both widths as 12.43 Hz. The answers are in hertz because we are using f_0 rather than ω_0.

If we desire a bandpass filter with transition widths less than some prescribed amount, we may solve for the maximum allowable TW in the corresponding low-pass case and use this value to select the appropriate low-pass filter. The bandpass function is then found from the low-pass data. Since by Eqs. (5-18) and (5-19) we have $TW_U > TW_L$, we shall take a given value of TW_U as the maximum allowable bandpass transition width. Then TW_L will also be less than this maximum allowable value. From Eq. (5-19), the corresponding normalized low-pass TW is

$$TW = \frac{2Q(TW_U)}{\omega_0}\left[\frac{Q(TW_U)/\omega_0 + \sqrt{1 + 4Q^2}}{1 + 2Q(TW_U)/\omega_0 + \sqrt{1 + 4Q^2}}\right] \quad (5\text{-}21)$$

As an example, suppose that we want an elliptic bandpass filter with $f_0 = 1000$ Hz, PRW $= 1$ dB, MSL $= 50$ dB, $Q = 10$, and transition widths of 10 Hz or less. Taking $TW_U = 10$ Hz, we have $Q(TW_U)/\omega_0 = 10(10)/1000 = 0.1$, and by Eq. (5-21),

$$TW = 2(0.1)\left[\frac{0.1 + \sqrt{401}}{1 + 0.2 + \sqrt{401}}\right]$$
$$= 0.1896$$

SEC. 5-4 INFINITE-GAIN MFB BANDPASS FILTERS

By Appendix D we must have an order of at least $N = 7$, for which $TW = 0.1013$ ($N = 6$ has $TW = 0.1989$, which is too large). For the case $N = 7$, we have, by Eqs. (5-18) and (5-19), the actual transition widths

$$TW_L = 4.8, \quad TW_U = 5.3 \text{ Hz}$$

both of which are less than the allowable value of 10.

5-4 INFINITE-GAIN MULTIPLE-FEEDBACK BANDPASS FILTERS

One of the simplest second-order bandpass circuits is the infinite-gain multiple-feedback (MFB) circuit of Fig. 5-7 [13].

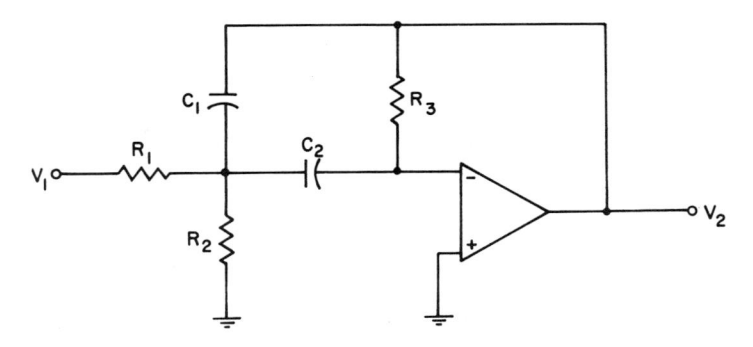

FIGURE 5-7 *Infinite-gain MFB bandpass filter.*

It realizes the bandpass function (5-11) with an inverting gain ($-\rho$ is obtained, for $\rho > 0$), where

$$\rho\omega_0 = \frac{1}{R_1 C_1}$$

$$\beta\omega_0 = \frac{1}{R_3}\left(\frac{1}{C_1} + \frac{1}{C_2}\right) \qquad (5\text{-}22)$$

$$\gamma\omega_0^2 = \frac{1}{R_3 C_1 C_2}\left(\frac{1}{R_1} + \frac{1}{R_2}\right)$$

For a given ω_0, ρ, β, and γ, the resistance values are given by

$$R_1 = \frac{1}{\rho\omega_0 C_1}$$

$$R_2 = \frac{\beta}{[C_1(\gamma - \rho\beta) + \gamma C_2]\omega_0} \qquad (5\text{-}23)$$

$$R_3 = \frac{1}{\beta\omega_0}\left(\frac{1}{C_1} + \frac{1}{C_2}\right)$$

where C_1 and C_2 are arbitrary. Thus we may select C_1 (preferably near $10/f_0$ μF) and C_2 so that $R_2 > 0$, and determine the resistances. The requirement on C_2 is

$$C_2 > \frac{C_1(\rho\beta - \gamma)}{\gamma} \qquad (5\text{-}24)$$

As an example, suppose that we want to construct a second-order bandpass filter with a center frequency $f_0 = 1000$ Hz, $Q = 5$, and a gain $K = 2$. The transfer function, given by Eq. (5-5), is

$$\frac{V_2}{V_1} = \frac{0.4\omega_0 s}{s^2 + 0.2\omega_0 s + \omega_0^2}$$

By comparison with Eq. (5-11) we see that $\rho = 0.4$, $\beta = 0.2$, and $\gamma = 1$. Choosing $C_1 = 10/f_0 = 0.01$ μF, we have, by Eq. (5-24),

$$C_2 > \frac{0.01[0.4(0.2) - 1]}{1} \mu F$$

Thus any positive C_2 will suffice. Choosing $C_2 = 0.01$ μF, then by Eq. (5-23), we have

$$R_1 = 39.79 \text{ k}\Omega$$
$$R_2 = 1.66 \text{ k}\Omega$$
$$R_3 = 159.15 \text{ k}\Omega$$

The MFB bandpass filter, like its low-pass and high-pass counterparts, has a minimal number of elements, an inverting gain, and is capable of Q's up to 10 for moderate gains. The design procedure is summarized in Section 5-10.

5-5 VCVS BANDPASS FILTERS

The VCVS circuit [18] of Fig. 5-8 realizes the second-order bandpass function (5-11), where

$$\rho\omega_0 = \frac{\mu}{R_1 C_1}$$

$$\beta\omega_0 = \frac{1}{C_1}\left[\frac{1}{R_1} + \frac{1}{R_2}(1-\mu) + \frac{2}{R_3}\right] \quad (5\text{-}25)$$

$$\gamma\omega_0^2 = \frac{1}{R_3 C_1^2}\left(\frac{1}{R_1} + \frac{1}{R_2}\right)$$

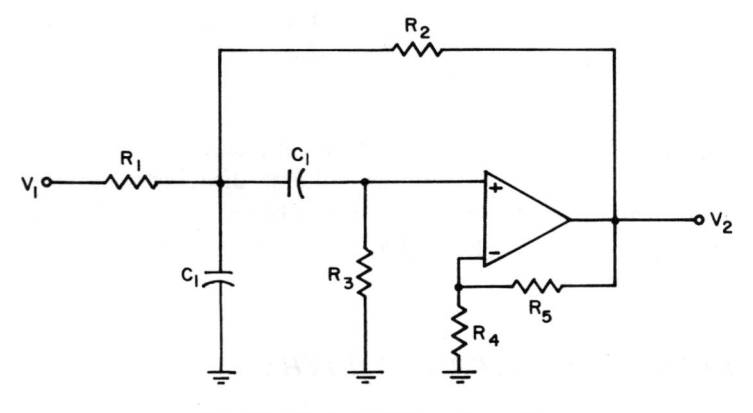

FIGURE 5-8 VCVS bandpass filter.

with

$$\mu = 1 + \frac{R_5}{R_4} \quad (5\text{-}26)$$

The resistance values for Fig. 5-8 are given by

$$R_1 = \frac{\mu}{\rho\omega_0 C_1}$$

$$R_2 = \frac{2(\mu-1)}{[\rho(2/\mu - 1) - \beta + \sqrt{(\rho-\beta)^2 + 8\gamma(\mu-1)}]\omega_0 C_1}$$

$$R_3 = \frac{1}{\gamma\omega_0^2 C_1^2}\left(\frac{1}{R_1} + \frac{1}{R_2}\right) \quad (5\text{-}27)$$

$$R_4 = \frac{\mu}{\mu - 1} R_3$$

$$R_5 = \mu R_3$$

where C_1 and μ are arbitrary, with

$$\mu = 1 + \frac{R_5}{R_4} > 1 \tag{5-28}$$

A considerable simplification is obtained if we choose $\mu = 2$, or equivalently, $R_4 = R_5$. In this case Eq. (5-27) becomes

$$R_1 = \frac{2}{\rho\omega_0 C_1}$$

$$R_2 = \frac{2}{[-\beta + \sqrt{(\rho - \beta)^2 + 8\gamma}]\omega_0 C_1}$$

$$R_3 = \frac{1}{\gamma\omega_0^2 C_1^2}\left(\frac{1}{R_1} + \frac{1}{R_2}\right) \tag{5-29}$$

$$R_4 = R_5 = 2R_3$$

The VCVS bandpass filter has the same advantages as the VCVS low-pass and high-pass filters, discussed previously. It yields a noninverting gain and can achieve a value of Q up to 10 for moderate gains. The design procedure is given in Section 5-11.

5-6 BIQUAD BANDPASS FILTERS

A biquad circuit [2] which realizes the second-order bandpass function of Eq. (5-11) is shown in Fig. 5-9. The circuit achieves

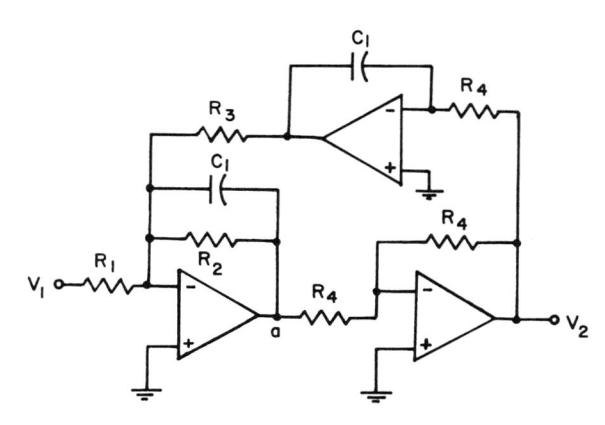

FIGURE 5-9 Biquad bandpass filter.

Eq. (5-11) with

$$p\omega_0 = \frac{1}{R_1 C_1}$$

$$\beta\omega_0 = \frac{1}{R_2 C_1} \qquad (5\text{-}30)$$

$$\gamma\omega_0^2 = \frac{1}{R_3 R_4 C_1^2}$$

The resistance values are given by

$$R_1 = \frac{1}{p\omega_0 C_1}$$

$$R_2 = \frac{1}{\beta\omega_0 C_1} = \frac{p}{\beta} R_1 \qquad (5\text{-}31)$$

$$R_3 = \frac{1}{\gamma\omega_0^2 C_1^2 R_4}$$

where C_1 and R_4 are arbitrary. The gain is noninverting, but an inverting gain may be obtained by taking the output at node a.

The biquad circuit requires more elements than the MFB or VCVS circuits, but it is a very popular design because of its stability and excellent tuning features. Also, it is capable of attaining values of Q up to 100. The design procedure is summarized in Section 5-12.

5-7 INVERSE CHEBYSHEV AND ELLIPTIC BANDPASS FILTERS

As was pointed out earlier, a second-order bandpass stage of an inverse Chebyshev or elliptic filter arising from a first-order section of the corresponding low-pass filter has the form of Eq. (5-11). Thus it may be realized by the methods of the previous three sections. Second-order sections arising from corresponding second-order low-pass sections occur in pairs having transfer functions given by Eqs. (5-12) and (5-13). These are special cases of the general function (5-15). This function is identical to the low-pass transfer function (3-16) of Section 3-3 with ω_c replaced by ω_0. Therefore, we may use the results and circuits of Sections

3-3 through 3-5, with the appropriate values of $p, \alpha, \beta,$ and γ, to realize the bandpass stage. This will be done in the design summaries given in Sections 5-13, 5-14, and 5-15.

5-8 TUNING THE SECOND-ORDER BANDPASS STAGES

The tuning of a second-order bandpass stage with transfer function

$$\frac{V_2}{V_1} = \frac{p\omega_0 s}{s^2 + \beta\omega_0 s + \gamma\omega_0^2} \qquad (5\text{-}32)$$

is more readily accomplished by observing the general shape that its amplitude response should take. This is shown in Fig. 5-10,

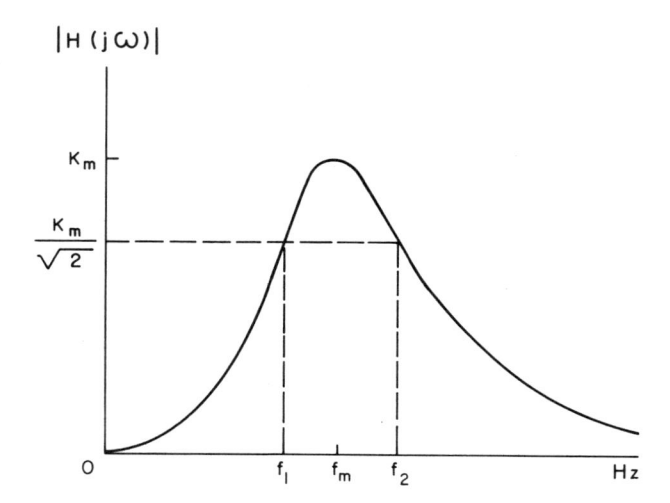

FIGURE 5-10 Second-order bandpass amplitude response.

where the peak value is

$$K_m = \frac{p}{\beta} \qquad (5\text{-}33)$$

occurring at

$$f_m = f_0 \sqrt{\gamma} \quad \text{Hz} \qquad (5\text{-}34)$$

with $f_0 = \omega_0/2\pi$.

SEC. 5-8 TUNING THE SECOND-ORDER BANDPASS   109

The frequencies f_1 and f_2 (Hz) are the 3-dB points and are given by

$$f_1 = f_0 \left[\frac{-\beta + \sqrt{\beta^2 + 4\gamma}}{2} \right]$$
$$f_2 = f_0 \left[\frac{\beta + \sqrt{\beta^2 + 4\gamma}}{2} \right] \tag{5-35}$$

The transfer function (5-32) is that of a second-order bandpass filter, where by Eq. (5-5) we have $\rho = K/Q$, $\beta = 1/Q$, and $\gamma = 1$. It is also the transfer function of a second-order stage of a higher-order bandpass filter, which arises from a corresponding first-order low-pass stage of Butterworth, Chebyshev, inverse Chebyshev, or elliptic type. In this case, by Eq. (5-4), we have $\rho = KC/Q$, $\beta = C/Q$, and $\gamma = 1$. Finally, Eq. (5-32) is the general form of the two bandpass stages arising from a corresponding second-order Butterworth or Chebyshev low-pass filter. In this case, by Eqs. (5-7) and (5-8) we have, for the two stages, $\rho = K_1 \sqrt{C}/Q$, $\beta = D/E$, $\gamma = D^2$, and $\rho = K_2 \sqrt{C}/Q$, $\beta = 1/DE$, $\gamma = 1/D^2$.

Second-order stages of elliptic and inverse Chebyshev bandpass filters corresponding to second-order low-pass stages have transfer functions of the form Eq. (5-15), which we repeat as

$$\frac{V_2}{V_1} = \frac{\rho(s^2 + \alpha\omega_0^2)}{s^2 + \beta\omega_0 s + \gamma\omega_0^2} \tag{5-36}$$

For each low-pass stage there will be two bandpass stages with functions like Eq. (5-36). These functions are given by Eqs. (5-12) and (5-13), from which we see that $\rho = K_1 \sqrt{C/A}$, $\alpha = A_1$, $\beta = D/E$, $\gamma = D^2$ in the first case and $\rho = K_2 \sqrt{C/A}$, $\alpha = 1/A_1$, $\beta = 1/DE$, $\gamma = 1/D^2$ in the second case. One of these is a low-pass stage with an amplitude response like Fig. 3-11(a) or (b). The other is a high-pass stage with an amplitude response like Fig. 4-7(a) or (b). The function (5-12) is the low-pass stage function if $A_1 > D^2$ and is the high-pass stage function otherwise.

In Figs. 3-11(a) and 4-7(a) the peak amplitude is

$$K_m = \frac{2\rho}{\beta} \sqrt{\frac{(\alpha - \gamma)^2 + \alpha\beta^2}{4\gamma - \beta^2}} \tag{5-37}$$

occurring at

$$f_m = f_0 \sqrt{\frac{2\gamma(\alpha - \gamma) - \alpha\beta^2}{2(\alpha - \gamma) + \beta^2}} \qquad (5\text{-}38)$$

The dc amplitude is $\rho\alpha/\gamma$ and $f_z = f_0\sqrt{\alpha}$ in all four cases of Figs. 3-11 and 4-7.

Tuning procedures for the individual bandpass circuits will be given in the design summaries to follow.

5-9 GENERAL DESIGN INFORMATION FOR BANDPASS FILTER CONSTRUCTION

In the design summaries to follow, circuits and short-cut procedures are given for obtaining second-order bandpass filters and second-order stages of higher-order Butterworth, Chebyshev, inverse Chebyshev, and elliptic bandpass filters. In every case the corresponding low-pass filter data of Appendix A, B, or C will be used to obtain the bandpass filters.

Second-order filters consist of one second-order stage which may be constructed using the MFB, VCVS, or biquad circuits of Section 5-10, 5-11, or 5-12.

In the case of higher-order filters, the number of second-order sections is the same as the order of the corresponding low-pass filter. The bandpass stage corresponding to a first-order low-pass stage may be constructed using the MFB, VCVS, or biquad circuits. There will be two second-order bandpass stages for each second-order low-pass stage. In the case of Butterworth and Chebyshev bandpass filters, these may also be MFB, VCVS, or biquad circuits, and in the case of inverse Chebyshev and elliptic bandpass filters, these may be constructed as summarized in Section 5-13, 5-14, or 5-15.

As an example, suppose that we wish to obtain a sixth-order inverse Chebyshev bandpass filter with a gain of 8, $f_0 = 1000$ Hz, $Q = 5$, and MSL $= 40$ dB. From Appendix B, for $N = 3$, we have a first-order low-pass stage with $C = 1.060226$ and a second-order low-pass stage with $A = 12.075684$, $B = 0.969938$, and $C = 1.028354$. Since there are three second-order bandpass stages, we shall take the gain per stage to be 2. The stage function corresponding to the first-order low-pass factor is given by Eq. (5-4), for which

$$\frac{KC\omega_0}{Q} = \frac{2(1.060226)(2000\pi)}{5} = 2664.639$$

$$\frac{C\omega_0}{Q} = 1332.319$$

$$\omega_0^2 = 39.478 \times 10^6$$

This function may be realized by the MFB, the VCVS, or the biquad circuit.

The two stage functions corresponding to the second-order low-pass factor are given by Eqs. (5-12) and (5-13), where $K_1 = K_2 = 2$. By Eqs. (5-14), (5-10), and (5-9) we have $A_1 = 1.977$, $D = 1.093$, and $E = 10.351$. Since $A_1 > D^2$, Eq. (5-12) is a low-pass function and Eq. (5-13) is a high-pass function. The two stages may be realized by the circuits of Section 3-3, 3-4, or 3-5.

5-10 MFB BANDPASS FILTER DESIGN SUMMARY

To design (a) a second-order bandpass filter, (b) a second-order stage of a higher-order Butterworth, Chebyshev, inverse Chebyshev, or elliptic bandpass filter corresponding to a first-order low-pass stage, or (c) a second-order stage of a higher-order Butterworth or Chebyshev bandpass filter corresponding to a second-order low-pass stage, having a given center frequency f_0 Hz (or $\omega_0 = 2\pi f_0$ rad/s), stage gain K, and quality factor Q, perform the following steps.

1. In case (b) find the normalized coefficient C of the first-order low-pass stage, and in case (c) find the normalized coefficients B and C of the second-order low-pass stage from the appropriate table of Appendix A, B, or C

2. Select a standard value of C_1 (preferably near $10/f_0$ μF) and a standard value of C_2 satisfying

$$C_2 > \frac{C_1(\rho\beta - \gamma)}{\gamma}$$

and calculate the resistance values given by

$$R_1 = \frac{1}{\rho\omega_0 C_1}$$

$$R_2 = \frac{\beta}{[C_1(\gamma - \rho\beta) + \gamma C_2]\omega_0}$$

$$R_3 = \frac{1}{\beta\omega_0}\left(\frac{1}{C_1} + \frac{1}{C_2}\right)$$

The values of ρ, β, and γ are given in case (a) by

$$\rho = \frac{K}{Q}$$

$$\beta = \frac{1}{Q}$$

$$\gamma = 1$$

and in case (b) by

$$\rho = \frac{KC}{Q}$$

$$\beta = \frac{C}{Q}$$

$$\gamma = 1$$

In case (c) there are two stages, one having

$$\rho = \frac{K\sqrt{C}}{Q}$$

$$\beta = \frac{D}{E}$$

$$\gamma = D^2$$

and the other having

$$\rho = \frac{K\sqrt{C}}{Q}$$

$$\beta = \frac{1}{DE}$$

$$\gamma = \frac{1}{D^2}$$

SEC. 5-10 MFB BANDPASS FILTER DESIGN

where

$$E = \frac{1}{B}\sqrt{\frac{C + 4Q^2 + \sqrt{(C + 4Q^2) - (2BQ)^2}}{2}}$$

$$D = \frac{1}{2}\left[\frac{BE}{Q} + \sqrt{\left(\frac{BE}{Q}\right)^2 - 4}\right]$$

3. Select standard values of the resistances as close as possible to the calculated values and construct the filter, or its stages, in accordance with Fig. 5-11. (See also Section 5-9 for general information on the various filter stages.)

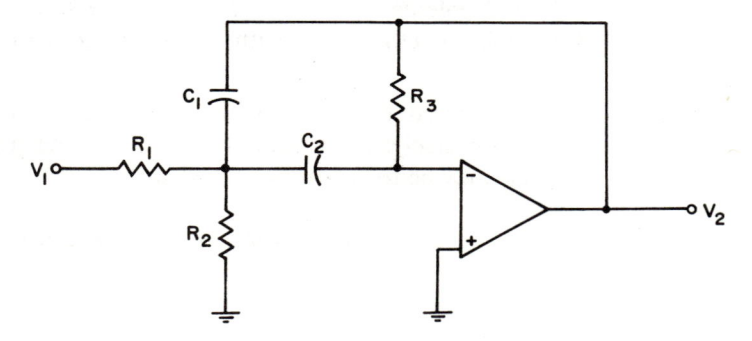

FIGURE 5-11 MFB bandpass filter circuit.

Comments

(a) For best performance, element values close to those selected and calculated should be used. Higher-order filters require more accurate element values than lower-order filters. The performance of the filter is unchanged if all the resistances are multiplied and the capacitances divided by a common factor

(b) The input impedance of the op amp should be at least $10R_{eq}$, where

$$R_{eq} = R_3$$

The open-loop gain of the op amp should be at least 50 times the amplitude of the filter, or stage, at f_a, the highest desired frequency in the passband, and its slew rate (volts per microsecond) should be at least $\frac{1}{2}\omega_a \times 10^{-6}$ times the peak-to-peak output voltage

(c) The gain is inverting with magnitude

$$K = \frac{R_3 C_2}{R_1(C_1 + C_2)}$$

Thus the gain can be adjusted by varying R_1. To adjust Q, R_2 may be varied, and to adjust the center frequency, both R_2 and R_3 may be varied simultaneously by the same percentage without affecting Q. These steps may be repeated if necessary. (See also Section 5-8.)

(d) The circuit should be used only for filter stages with gain $K = \rho/\beta$ and pole-pair $Q = \sqrt{\gamma}/\beta$ of 10 or less. The gain could be higher if Q is lower, with limits of $KQ = 100$ and $Q = 10$

(e) The order required for given upper and lower transition widths or, conversely, the transition widths ensuing from a given order, may be found from Section 5-3

The MFB bandpass filter was discussed and an example given in Section 5-4.

5-11 VCVS BANDPASS FILTER DESIGN SUMMARY

To design (a) a second-order bandpass filter, (b) a second-order stage of a higher-order Butterworth, Chebyshev, inverse Chebyshev, or elliptic bandpass filter corresponding to a first-order low-pass stage, or (c) a second-order stage of a higher-order Butterworth or Chebyshev bandpass filter corresponding to a second-order low-pass stage, having a given center frequency f_0 Hz (or $\omega_0 = 2\pi f_0$ rad/s), stage gain K, and quality factor Q, perform the following steps.

1. In case (b) find the normalized coefficient C of the first-order low-pass stage, and in case (c) find the normalized coefficients B and C of the second-order low-pass stage from the appropriate table of Appendix A, B, or C

2. Select a standard value of C_1 (preferably near $10/f_0$ μF) and

SEC. 5-11 VCVS BANDPASS FILTER DESIGN

calculate the resistance values given by

$$R_1 = \frac{2}{\rho \omega_0 C_1}$$

$$R_2 = \frac{2}{[-\beta + \sqrt{(\rho - \beta)^2 + 8\gamma}]\omega_0 C_1}$$

$$R_3 = \frac{1}{\gamma \omega_0^2 C_1^2}\left(\frac{1}{R_1} + \frac{1}{R_2}\right)$$

$$R_4 = R_5 = 2R_3$$

(A more general design with $R_4 \neq R_5$ is given in Section 5-5.) The values of ρ, β, and γ are given in case (a) by

$$\rho = \frac{K}{Q}$$

$$\beta = \frac{1}{Q}$$

$$\gamma = 1$$

and in case (b) by

$$\rho = \frac{KC}{Q}$$

$$\beta = \frac{C}{Q}$$

$$\gamma = 1$$

In case (c) there are two stages, one having

$$\rho = \frac{K\sqrt{C}}{Q}$$

$$\beta = \frac{D}{E}$$

$$\gamma = D^2$$

and the other having

$$\rho = \frac{K\sqrt{C}}{Q}$$

$$\beta = \frac{1}{DE}$$

$$\gamma = \frac{1}{D^2}$$

where

$$E = \frac{1}{B}\sqrt{\frac{C + 4Q^2 + \sqrt{(C + 4Q^2)^2 - (2BQ)^2}}{2}}$$

$$D = \frac{1}{2}\left[\frac{BE}{Q} + \sqrt{\left(\frac{BE}{Q}\right)^2 - 4}\right]$$

3. Select standard values of the resistances as close as possible to the calculated values and construct the filter, or its stages, in accordance with Fig. 5-12. (See also Section 5-9 for general information on the various filter stages.)

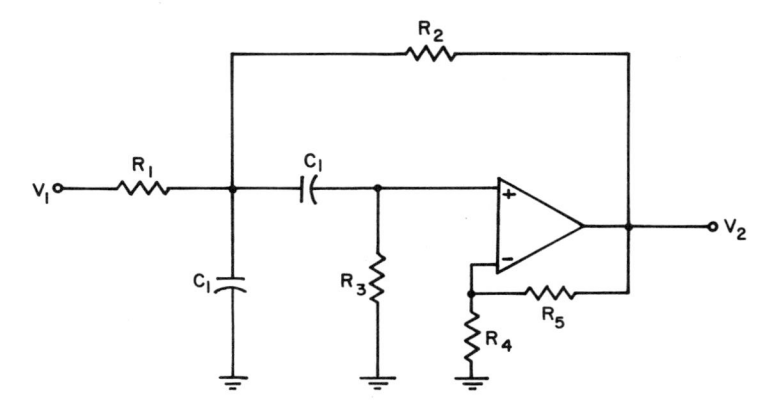

FIGURE 5-12 VCVS bandpass filter circuit.

Comments

(a) Comments (a), (b), (d), and (e) for the MFB filter of Section 5-10 apply directly

(b) The values of R_4 and R_5 are chosen to minimize the dc offset of the op amp. Other values may be used as long as the ratio $R_5/R_4 = 1$ (or $1 + R_5/R_4 = \mu$, using the more general design of Section 5-5)

(c) The gain of the filter is a noninverting one and may be adjusted by varying R_1 (or, in the more general design of Section 5-5, by varying the ratio R_5/R_4). The center frequency may be adjusted by varying R_3 and Q may be adjusted by

SEC. 5-12   BIQUAD BANDPASS FILTER DESIGN   117

varying R_2 and R_3. These steps affect each other, however, and may have to be repeated. (See Section 5-8 for other tuning features.)

The VCVS bandpass filter was discussed in Section 5-5.

5-12 BIQUAD BANDPASS FILTER DESIGN SUMMARY

To design (a) a second-order bandpass filter, (b) a second-order stage of a higher-order Butterworth, Chebyshev, inverse Chebyshev, or elliptic bandpass filter corresponding to a first-order low-pass stage, or (c) a second-order stage of a higher-order Butterworth or Chebyshev bandpass filter corresponding to a second-order low-pass stage, having a given center frequency f_0 Hz (or $\omega_0 = 2\pi f_0$ rad/s), stage gain K, and quality factor Q, perform the following steps.

1. In case (b) find the normalized coefficient C of the first-order low-pass stage, and in case (c) find the normalized coefficients B and C of the second-order low-pass stage from the appropriate table of Appendix A, B, or C

2. Select a standard value of C_1 (preferably near $10/f_0$ μF) and calculate the resistance values given by

$$R_1 = \frac{1}{\rho \omega_0 C_1}$$

$$R_2 = \frac{1}{\beta \omega_0 C_1} = \frac{\rho}{\beta} R_1$$

$$R_3 = \frac{1}{\gamma \omega_0^2 C_1^2 R_4}$$

where R_4 is arbitrary. A reasonable value is

$$R_4 = R_3 = \frac{1}{\sqrt{\gamma}\, \omega_0 C_1}$$

The values of p, β, and γ are given in case (a) by

$$p = \frac{K}{Q}$$

$$\beta = \frac{1}{Q}$$

$$\gamma = 1$$

and in case (b) by

$$p = \frac{KC}{Q}$$

$$\beta = \frac{C}{Q}$$

$$\gamma = 1$$

In case (c) there are two stages, one having

$$p = \frac{K\sqrt{C}}{Q}$$

$$\beta = \frac{D}{E}$$

$$\gamma = D^2$$

and the other having

$$p = \frac{K\sqrt{C}}{Q}$$

$$\beta = \frac{1}{DE}$$

$$\gamma = \frac{1}{D^2}$$

where

$$E = \frac{1}{B}\sqrt{\frac{C + 4Q^2 + \sqrt{(C + 4Q^2)^2 - (2BQ)^2}}{2}}$$

$$D = \frac{1}{2}\left[\frac{BE}{Q} + \sqrt{\left(\frac{BE}{Q}\right)^2 - 4}\right]$$

3. Select standard values of the resistances as close as possible to the calculated values and construct the filter, or its stages, in accordance with Fig. 5-13. (See also Section 5-9 for general information on the various filter stages.)

SEC. 5-13   VCVS BANDPASS ELLIPTIC FILTER DESIGN

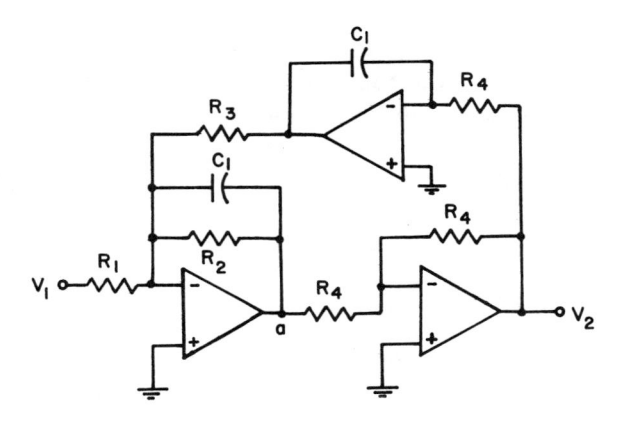

FIGURE 5-13 Biquad bandpass filter.

Comments

(a) Comments (a), (b), and (e) for the MFB filter of Section 5-10 apply directly, except that in (b), R_{eq} for each op amp is the resistance R_1 or R_4 connected to its inverting input terminal

(b) The stage gain is noninverting, given by $K = R_2/R_1$. If an inverting gain is desired, the output may be taken at node a

(c) Tuning may be accomplished by varying R_3 to adjust the center frequency, varying R_2 to adjust the pole-pair $Q = \sqrt{\gamma/\beta}$, and varying R_1 to adjust the gain. The value of R_4 may be chosen arbitrarily to maintain a smaller spread of resistance values. (See Section 5-8 for other tuning features.)

(d) The circuit may be used for pole-pair Q values up to 100

The biquad bandpass filter was discussed in Section 5-6.

5-13 VCVS BANDPASS ELLIPTIC FILTER DESIGN SUMMARY

To design the two second-order bandpass stages corresponding to a second-order low-pass stage of a higher-order inverse Chebyshev or elliptic bandpass fiter, having a given center frequency f_0 Hz (or $\omega_0 = 2\pi f_0$ rad/s), stage gain K, and quality factor Q, perform the following steps. (If the low-pass filter has a first-order stage, the corresponding second-order bandpass stage may be constructed using the circuits of Section 5-10, 5-11, or 5-12, as discussed in Section 5-9.)

1. Find the normalized coefficients A, B, and C of the corresponding second-order low-pass stage from the appropriate table of Appendix B or C

2. Select a standard value of C_1 (preferably near $10/f_0$ μF) and calculate the resistance values. The first-stage values are

$$R_1 = \frac{\mu D}{KA_1 E\omega_0 C_1}\sqrt{\frac{A}{C}}$$

$$R_2 = \frac{E}{D\omega_0 C_2}$$

$$R_3 = \frac{\mu}{DE\omega_0 C_1}$$

$$R_4 = \frac{K}{\mu}\sqrt{\frac{C}{A}}R_5$$

$$R_6 = \frac{\mu R_2}{\mu - 1}$$

$$R_7 = \mu R_2$$

The second-stage values are

$$R_1 = \frac{\mu A_1}{KDE\omega_0 C_1}\sqrt{\frac{A}{C}}$$

$$R_2 = \frac{DE}{\omega_0 C_2}$$

$$R_3 = \frac{\mu D}{E\omega_0 C_1}$$

$$R_4 = \frac{K}{\mu}\sqrt{\frac{C}{A}}R_5$$

$$R_6 = \frac{\mu R_2}{\mu - 1}$$

$$R_7 = \mu R_2$$

SEC. 5-13 VCVS BANDPASS ELLIPTIC FILTER DESIGN 121

In both cases C_2, R_5, and $\mu > 1$ are arbitrary, and A_1, E, and D are given by

$$A_1 = 1 + \frac{1}{2Q^2}(A + \sqrt{A^2 + 4AQ^2})$$

$$E = \frac{1}{B}\sqrt{\frac{C + 4Q^2 + \sqrt{(C + 4Q^2)^2 - (2BQ)^2}}{2}}$$

$$D = \frac{1}{2}\left[\frac{BE}{Q} + \sqrt{\left(\frac{BE}{Q}\right)^2 - 4}\right]$$

If K and the pole-pair $Q = E$ are moderate values, such as 10 or less, reasonable values of the arbitrary parameters are

$$C_2 = C_1$$

$$R_5 = \frac{1}{\omega_0 C_1}$$

and $\mu = 2$ (in which case $R_6 = R_7$). If K and/or E are high, say over 10, then C_2, R_5, and μ should be chosen to maintain a low spread of resistance values

3. Select standard values of resistances as close as possible to the calculated values and construct the filter stages in accordance with Fig. 5-14

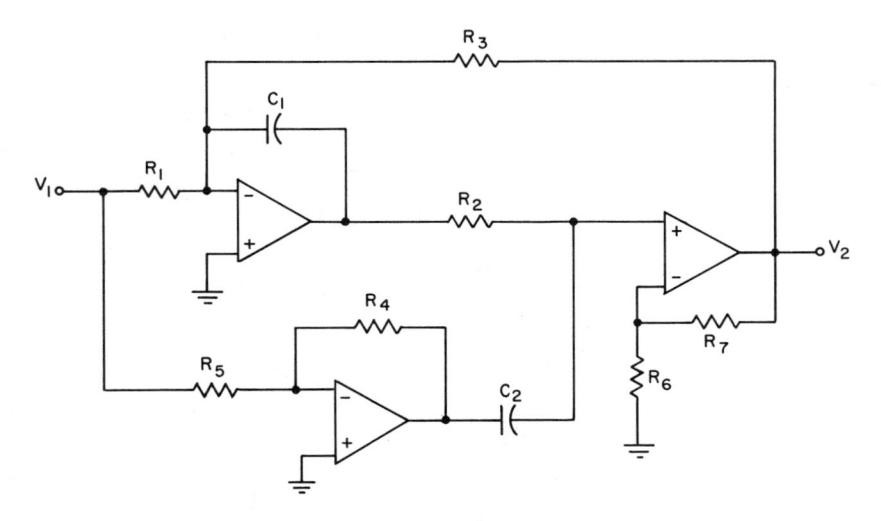

FIGURE 5-14 VCVS bandpass elliptic filter circuit.

4. If $\mu = 1$ is desired, R_6 becomes an open circuit and R_7 becomes a short circuit, with the other resistances given as in step 2. In this case the circuit is the voltage follower circuit of Fig. 3-6.

Comments

(a) Comments (a), (b), and (e) for the MFB filter of Section 5-10 apply directly, except that in (b), R_{eq} for each op amp is the resistance R_1, R_2, or R_5 connected to its inverting input terminal

(b) The circuit may be used for both high and low pole-pair $Q = E$ by selecting the arbitrary parameters of step 2 to keep the spread of resistance values relatively low. For moderate gains, E may be as high as 100

(c) Tuning may be accomplished by varying the ratio R_4/R_5 to move the notch toward f_z, shown in Fig. 3-11 for the low-pass section ($A_1 > D^2$) and in Fig. 4-7 for the high-pass section ($A_1 < D^2$). The gain μ of the VCVS, given by

$$\mu = 1 + \frac{R_7}{R_6}$$

then may be adjusted, by varying the ratio R_7/R_6, to move the peak value K_m toward f_m, where for the first stage,

$$K_m = \frac{2KE}{D^2}\sqrt{\frac{C}{A}\left[\frac{E^2(A_1 - D^2)^2 + A_1 D^2}{4E^2 - 1}\right]}$$

$$f_m = f_0\sqrt{\frac{D^2[2E^2(A_1 - D^2) - A_1]}{2E^2(A_1 - D^2) + D^2}}$$

and for the second stage,

$$K_m = \frac{2KE}{A_1}\sqrt{\frac{C}{A}\left[\frac{E^2(D^2 - A_1)^2 + A_1 D^2}{4E^2 - 1}\right]}$$

$$f_m = \frac{f_0}{D}\sqrt{\frac{2E^2(D^2 - A_1) - D^2}{2E^2(D^2 - A_1) + A_1}}$$

SEC. 5-14   3-CAPACITOR BP FILTER DESIGN   **123**

Finally, R_2 may be varied to obtain K_m. These steps may be repeated until the stages are tuned (see also Section 5-8)

(d) The circuit yields an inverting gain $-K$ ($K > 0$).

The VCVS bandpass elliptic filter was discussed in Section 5-7.

5-14 THREE-CAPACITOR BANDPASS ELLIPTIC FILTER DESIGN SUMMARY

To design the two second-order bandpass stages corresponding to a second-order low-pass stage of a higher-order inverse Chebyshev or elliptic bandpass filter, having a given center frequency f_0 Hz (or $\omega_0 = 2\pi f_0$ rad/s), stage gain K, and quality factor Q, perform the following steps. (If the low-pass filter has a first-order stage, the corresponding second-order bandpass stage may be constructed using the circuits of Section 5-10, 5-11, or 5-12, as discussed in Section 5-9.)

1. Find the normalized coefficients A, B, and C of the corresponding second-order low-pass stage from the appropriate table of Appendix B or C

2. Select a standard value of C_1 (preferably near $10/f_0$ μF) and calculate the element values. The first-stage values are

$$C_3 = K\sqrt{\frac{C}{A}}\, C_2$$

$$R_1 = \frac{1}{R_4 A_1 \omega_0^2 C_1 C_3}$$

$$R_2 = \frac{1}{R_4 D^2 \omega_0^2 C_1 C_2} = \frac{KA_1}{D^2}\sqrt{\frac{C}{A}}\, R_1$$

$$R_3 = \frac{E}{D\omega_0 C_2}$$

The second-stage values are

$$C_3 = K\sqrt{\frac{C}{A}}\, C_2$$

$$R_1 = \frac{A_1}{R_4 \omega_0^2 C_1 C_3}$$

$$R_2 = \frac{D^2}{R_4 \omega_0^2 C_1 C_2} = \frac{KD^2}{A_1}\sqrt{\frac{C}{A}}\, R_1$$

$$R_3 = \frac{DE}{\omega_0 C_2}$$

In both cases C_2 and R_4 are arbitrary, and A_1, E, and D are given by

$$A_1 = 1 + \frac{1}{2Q^2}(A + \sqrt{A^2 + 4AQ^2})$$

$$E = \frac{1}{B}\sqrt{\frac{C + 4Q^2 + \sqrt{(C + 4Q^2)^2 - (2BQ)^2}}{2}}$$

$$D = \frac{1}{2}\left[\frac{BE}{Q} + \sqrt{\left(\frac{BE}{Q}\right)^2 - 4}\right]$$

If the pole-pair $Q = E$ is low, C_2 may be chosen near C_1, and if E is high, C_2 may be chosen larger than C_1. In either case, R_4 may be selected to minimize the spread of the resistance values

3. Select standard values of resistances and capacitances as close as possible to the calculated values and construct the filter stages in accordance with Fig. 5-15

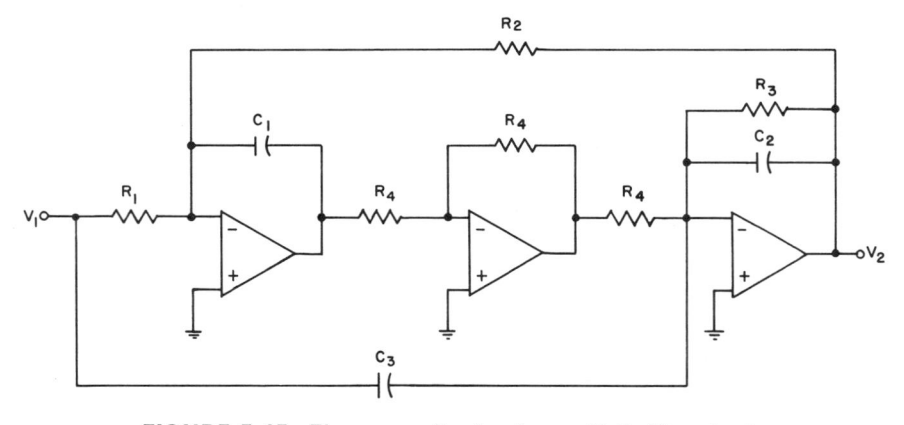

FIGURE 5-15 Three-capacitor bandpass elliptic filter circuit.

SEC. 5-15   BIQUAD BP ELLIPTIC FILTER DESIGN   125

Comments

(a) Comments (a), (b), and (e) for the MFB filter of Section 5-10 apply directly, except that in (b), R_{eq} for each op amp is the resistance R_1 or R_4 connected to its inverting input

(b) The circuit may be used for both high and low pole-pair $Q = E$, with an upper limit of approximately $E = 100$

(c) Tuning may be accomplished by first adjusting R_1 until the notch is at f_z, shown in Fig. 3-11 for the low-pass section ($A_1 > D^2$) and in Fig. 4-7 for the high-pass section ($A_1 < D^2$). The peak amplitude K_m may be moved to f_m by adjusting R_2; the values of K_m and f_m for the two stages are given in comment (c) of the VCVS bandpass elliptic filter summary in Section 5-13. Finally, R_3 may be adjusted to obtain the correct value of K_m. For low $Q = E$ sections the last two steps may need to be repeated

(d) The circuit yields an inverting gain $-K$ ($K > 0$)

The three-capacitor circuit was discussed in Section 5-7.

5-15 BIQUAD BANDPASS ELLIPTIC FILTER DESIGN SUMMARY

To design the two second-order bandpass stages corresponding to a second-order low-pass stage of a higher-order inverse Chebyshev or elliptic bandpass filter, having a given center frequency f_0 Hz (or $\omega_0 = 2\pi f_0$ rad/s), stage gain K, and quality factor Q, perform the following steps. (If the low-pass filter has a first-order stage, the corresponding second-order bandpass stage may be constructed using the circuits of Section 5-10, 5-11, or 5-12, as discussed in Section 5-9.)

1. Find the normalized coefficients A, B, and C of the corresponding second-order low-pass stage from the appropriate table of Appendix B or C

2. Select a standard value of C_1 (preferably near $10/f_0$ μF) and

calculate the element values. The first-stage values are

$$R_1 = \frac{E}{KD\omega_0 C_1}\sqrt{\frac{A}{C}}$$

$$R_2 = K\sqrt{\frac{C}{A}}R_1$$

$$R_3 = \frac{1}{D\omega_0 C_1}$$

$$R_4 = \sqrt{\frac{A}{C}}\frac{R_7}{K}$$

$$R_5 = \frac{D}{KA_1\omega_0 C_2}\sqrt{\frac{A}{C}}$$

$$R_6 = \frac{C_1 R_3}{C_2}$$

The second-stage values are

$$R_1 = \frac{DE}{K\omega_0 C_1}\sqrt{\frac{A}{C}}$$

$$R_2 = K\sqrt{\frac{C}{A}}R_1$$

$$R_3 = \frac{D}{\omega_0 C_1}$$

$$R_4 = \sqrt{\frac{A}{C}}\frac{R_7}{K}$$

$$R_5 = \frac{A_1}{KD\omega_0 C_2}\sqrt{\frac{A}{C}}$$

$$R_6 = \frac{C_1 R_3}{C_2}$$

In both cases C_2 and R_7 are arbitrary, and A_1, E, and D are given by

$$A_1 = 1 + \frac{1}{2Q^2}(A + \sqrt{A^2 + 4AQ^2})$$

$$E = \frac{1}{B}\sqrt{\frac{C + 4Q^2 + \sqrt{(C + 4Q^2)^2 - (2BQ)^2}}{2}}$$

$$D = \frac{1}{2}\left[\frac{BE}{Q} + \sqrt{\left(\frac{BE}{Q}\right)^2 - 4}\right]$$

SEC. 5-15 BIQUAD BP ELLIPTIC FILTER DESIGN

Depending on the gain K and the pole-pair $Q = E$, C_2 and R_7 may be selected to minimize the spread of the resistance values

3. Select standard values of resistances and capacitances as close as possible to the calculated values and construct the filter stages in accordance with Fig. 5-16

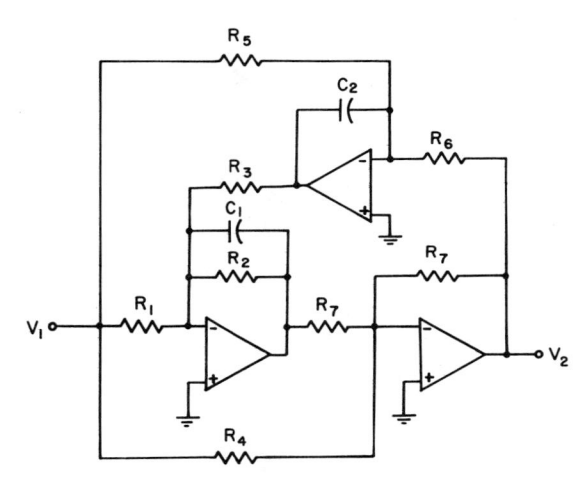

FIGURE 5-16 Biquad bandpass elliptic filter circuit.

Comments

(a) Comments (a), (b), and (e) for the MFB filter of Section 5-10 apply directly, except that in (b), R_{eq} for each op amp is the resistance R_1, R_6, or R_7 connected to its inverting input

(b) The circuit may be used for both high and low pole-pair $Q = E$, with an upper limit of approximately $E = 100$

(c) Tuning may be accomplished by adjusting R_4 to set the notch at f_z, adjusting R_3 to set the center frequency, adjusting R_2 to set Q, and adjusting R_1 or R_5 to set the gain

(d) The circuit yields an inverting gain $-K$ ($K > 0$), which is proportional to R_2/R_1

The biquad elliptic circuit was discussed in Section 5-7.

6

BAND-REJECT FILTERS

6-1 THE GENERAL CASE

A *band-reject filter* (also called *band-stop*, or *band-elimination*, or *notch*) is one that rejects a single band of frequencies and passes all other frequencies. The band rejected is of *bandwidth* BW centered approximately about a *center frequency* ω_0 rad/s, or $f_0 = \omega_0/2\pi$ Hz. An ideal and a practical band-reject amplitude response are shown in Fig. 6-1. In the practical case, shown with a solid line, the frequencies ω_L and ω_U are the *lower* and *upper cutoff frequencies*, which define the rejected band $\omega_L \leq \omega \leq \omega_U$ and the bandwidth $\text{BW} = \omega_U - \omega_L$. All these quantities have their counterparts in the bandpass filter, discussed in Chapter 5.

With reference to the practical response of Fig. 6-1, in the rejected band the amplitude never exceeds some prescribed value, such as A_1. Also, there are two passbands, $0 \leq \omega \leq \omega_L$ and $\omega \geq$

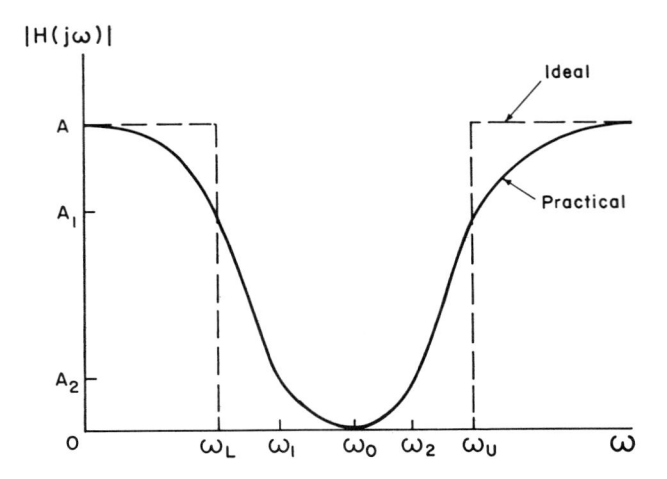

FIGURE 6-1 Ideal and practical band-reject response.

ω_U, where the amplitude is always greater than A_1. To be consistent with our discussions of filters in the previous chapters, we shall define a *stopband* $\omega_1 \leq \omega \leq \omega_2$, where the amplitude never exceeds a prescribed value $A_2 < A_1$. Then the regions $\omega_L < \omega < \omega_1$ and $\omega_2 < \omega < \omega_U$ are, respectively, the *lower* and *upper transition* bands, in which the response is monotonic.

The ratio $Q = \omega_0/\text{BW}$, like its bandpass counterpart, is the *quality factor* of the filter, which determines its selectivity. High Q corresponds to a relatively narrow bandwidth and low Q corresponds to a relatively wide bandwidth. The *gain K* of the filter is the dc value of the amplitude; that is, $K = |H(j0)|$.

Band-reject transfer functions may be obtained from normalized low-pass functions of S by the transformation [16]

$$S = \frac{\text{BW} \cdot s}{s^2 + \omega_0^2} = \frac{\omega_0 s}{Q(s^2 + \omega_0^2)} \qquad (6\text{-}1)$$

Consequently, the band-reject filter, like the bandpass filter, is always of even order, $n = 2, 4, 6, \ldots$. The resulting band-reject filter will be of Butterworth, Chebyshev, inverse Chebyshev, or elliptic type, depending on its corresponding low-pass type. A

Butterworth band-reject filter amplitude varies monotonically on either side of its notch, or center, frequency, as in the case of Fig. 6-1. A Chebyshev band-reject filter has passband ripples, an inverse Chebyshev band-reject filter has stopband ripples, and an elliptic band-reject filter has both pass- and stopband ripples. In every case the center and cutoff frequencies are related by $\omega_0 = \sqrt{\omega_L \omega_U}$.

The frequencies ω_L and ω_U of Fig. 6-1, which determine the two passbands, are given by

$$\omega_L = \omega_0 \left(-\frac{1}{2Q} + \sqrt{1 + \frac{1}{4Q^2}} \right)$$
$$\omega_U = \omega_0 \left(\frac{1}{2Q} + \sqrt{1 + \frac{1}{4Q^2}} \right) \tag{6-2}$$

The stopband frequencies ω_1 and ω_2 are given by

$$\omega_1 = \omega_0 \left(-\frac{1}{2\Omega_s Q} + \sqrt{1 + \frac{1}{4\Omega_s^2 Q^2}} \right)$$
$$\omega_2 = \omega_0 \left(\frac{1}{2\Omega_s Q} + \sqrt{1 + \frac{1}{4\Omega_s^2 Q^2}} \right) \tag{6-3}$$

where Ω_s is the beginning of the stopband of the corresponding low-pass filter. In other words,

$$\Omega_s = 1 + \text{TW} \tag{6-4}$$

where TW is the normalized transition width of the corresponding low-pass filter, given previously in Eqs. (2-16), (2-17), and (3-7), for $\omega_c = 1$, in the Butterworth, Chebyshev, and inverse Chebyshev cases, and tabulated in Appendix C for the elliptic case. We may note also that $\omega_0 = \sqrt{\omega_1 \omega_2}$.

Examples of band-reject responses of actual circuits are shown in Figs. 6-2 and 6-3. Figure 6-2 is that of a fourth-order 1-dB Chebyshev band-reject filter with $f_0 = 60$ Hz and $Q = 10$. Figure 6-3 is a sixth-order elliptic band-reject response with a 3-dB passband ripple width, a minimum stopband loss of 40 dB, $f_0 = 60$ Hz, and $Q = 5$.

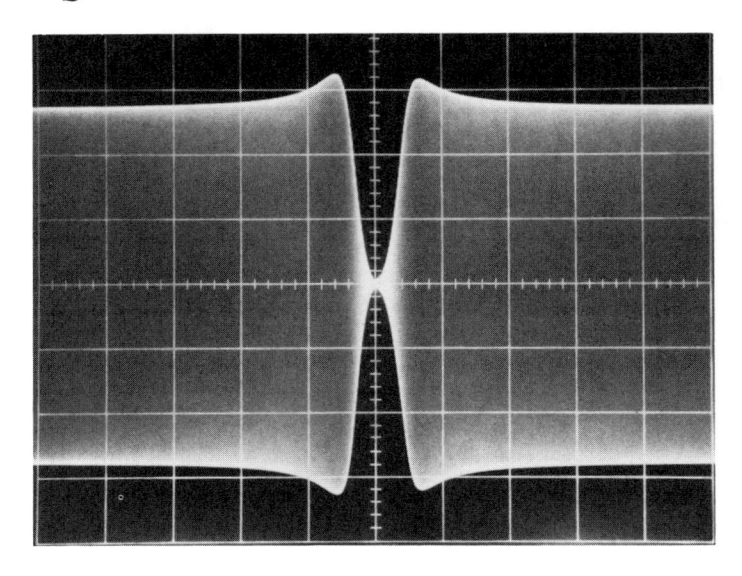

FIGURE 6-2 Fourth-order Chebyshev band-reject response.

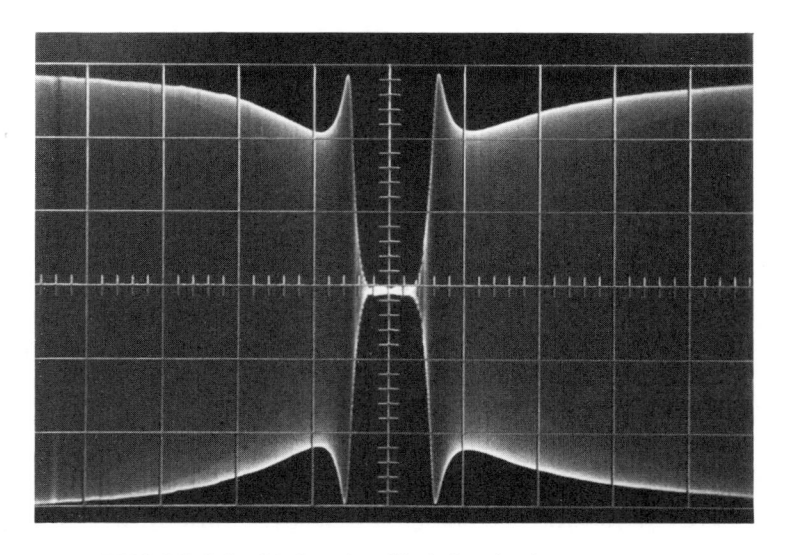

FIGURE 6-3 Sixth-order elliptic band-reject response.

6-2 TRANSFER FUNCTIONS

Like the bandpass functions of Chapter 5, band-reject functions are also obtained from corresponding low-pass functions. The band-reject function is a product of factors, each of which arises from a low-pass factor. In the case of a first-order low-pass factor,

$$\frac{V_2}{V_1} = \frac{KC}{S+C} \qquad (6\text{-}5)$$

the corresponding band-reject factor is the second-order function

$$\frac{V_2}{V_1} = \frac{K(s^2 + \omega_0^2)}{s^2 + (\omega_0/CQ)s + \omega_0^2} \qquad (6\text{-}6)$$

where C is the normalized coefficient of the corresponding low-pass first-order stage, given in Appendix A for the Butterworth and Chebyshev filters, in Appendix B for the inverse Chebyshev filter, and in Appendix C for the elliptic filter.

A second-order band-reject filter results when the corresponding low-pass filter is of first order. In this case Eq. (6-5) with $C = 1$ is the low-pass function, and by Eq. (6-6) we have

$$\frac{V_2}{V_1} = \frac{K(s^2 + \omega_0^2)}{s^2 + (\omega_0/Q)s + \omega_0^2} \qquad (6\text{-}7)$$

which is the transfer function of a second-order band-reject filter. This function could be thought of as that associated with a second-order Butterworth or Chebyshev band-reject filter, but these adjectives are generally reserved for higher-order band-reject filters.

Butterworth or Chebyshev band-reject transfer-function factors arising from second-order low-pass stages are of the form

$$\frac{V_2}{V_1} = \frac{K(s^2 + \omega_0^2)^2}{s^4 + (B\omega_0/CQ)s^3 + (2 + 1/CQ^2)\omega_0^2 s^2 + (B\omega_0^3/CQ)s + \omega_0^4}$$
$$(6\text{-}8)$$

where B and C are the corresponding low-pass coefficients of Appendix A. In Eq. (6-6) K is the stage gain and in Eq. (6-8) K is the overall gain of two second-order stages cascaded to realize the fourth-order function.

The transfer function (6-8) may be factored into the two second-order functions [4]

$$\left(\frac{V_2}{V_1}\right)_1 = \frac{K_1(s^2 + \omega_0^2)}{s^2 + (D_1\omega_0/E_1)s + D_1^2\omega_0^2} \qquad (6\text{-}9)$$

and

$$\left(\frac{V_2}{V_1}\right)_2 = \frac{K_2(s^2 + \omega_0^2)}{s^2 + (\omega_0/D_1 E_1)s + \omega_0^2/D_1^2} \qquad (6\text{-}10)$$

where

$$E_1 = \frac{1}{B}\sqrt{\frac{C}{2}[1 + 4CQ^2 + \sqrt{(1 + 4CQ^2)^2 - (2BQ)^2}]} \qquad (6\text{-}11)$$

and

$$D_1 = \frac{1}{2}\left[\frac{BE_1}{QC} + \sqrt{\left(\frac{BE_1}{QC}\right)^2 - 4}\right] \qquad (6\text{-}12)$$

Thus the transfer function of a Butterworth or Chebyshev band-reject filter of order $n = 4, 6, 8, \ldots$, will have a factor like Eq. (6-9) and a factor like Eq. (6-10) for each second-order stage in its corresponding low-pass filter. The numbers K_1 and K_2 are the gains of the two band-reject stages and should be chosen so that $K_1 K_2 = K$.

In summary, a typical transfer function of a second-order band-reject filter, or a second-order stage of a higher-order Butterworth or Chebyshev band-reject filter, is of the form

$$\frac{V_2}{V_1} = \frac{p(s^2 + \omega_0^2)}{s^2 + \beta\omega_0 s + \gamma\omega_0^2} \qquad (6\text{-}13)$$

where p, β, and γ are obtained by matching Eq. (6-13) with the appropriate one of Eq. (6-6), (6-7), (6-9), or (6-10).

In both Eqs. (6-9) and (6-10) E_1 is the pole-pair Q of each stage, and as in the filters of the previous chapters, high Q usually requires more elaborate circuits.

In the case of elliptic or inverse Chebyshev band-reject filters, the transfer function may also be factored into second-order functions. If the corresponding low-pass filter is of odd order, the band-reject factor arising from the first-order low-pass stage is

Eq. (6-6), which is a special case of Eq. (6-13), as already noted. The two factors arising from each second-order low-pass stage have the forms

$$\left(\frac{V_2}{V_1}\right)_1 = \frac{K_1(s^2 + A_2\omega_0^2)}{s^2 + (D_1\omega_0/E_1)s + D_1^2\omega_0^2} \quad (6\text{-}14)$$

and

$$\left(\frac{V_2}{V_1}\right)_2 = \frac{K_2(s^2 + \omega_0/A_2)}{s^2 + (\omega_0/D_1E_1)s + \omega_0^2/D_1^2} \quad (6\text{-}15)$$

where E_1 and D_1 are given by Eqs. (6-11) and (6-12) and

$$A_2 = 1 + \frac{1}{2AQ^2}(1 + \sqrt{1 + 4AQ^2}) \quad (6\text{-}16)$$

The coefficients A, B, and C are those of the normalized low-pass functions of Appendix B or C, and K_1 and K_2 are the stage gains. In general, Eqs. (6-14) and (6-15) are of the form

$$\frac{V_2}{V_1} = \frac{\rho(s^2 + \alpha\omega_0^2)}{s^2 + \beta\omega_0 s + \gamma\omega_0^2} \quad (6\text{-}17)$$

Indeed, Eq. (6-13) is simply Eq. (6-17) with $\alpha = 1$. Therefore, we may say that Eq. (6-17) includes all the band-reject stages discussed in this chapter. It should be noted that Eq. (6-17) is identical to Eq. (3-16), the low-pass elliptic and inverse Chebyshev function, when ω_c is replaced by ω_o.

6-3 TRANSITION WIDTHS

As discussed in Section 6-1, a band-reject filter has two passbands, $0 \leq \omega \leq \omega_L$ and $\omega \geq \omega_U$, where ω_L and ω_U are the lower and upper cutoff frequencies. Between the stopband, $\omega_1 \leq \omega \leq \omega_2$, and each of these passbands is a transition band. The lower transition band, $\omega_L \leq \omega \leq \omega_1$, has transition width

$$TW_L = \omega_1 - \omega_L \quad (6\text{-}18)$$

and the upper transition band, $\omega_2 \leq \omega \leq \omega_U$, has width

$$TW_U = \omega_U - \omega_2 \quad (6\text{-}19)$$

The low-pass to band-reject transformation (6-1) may also be used to relate the transition widths of the band-reject filter to that of its corresponding low-pass filter. These relations are

$$\frac{TW_L}{\omega_0} = \frac{TW + [\sqrt{1 + 4(TW + 1)^2 Q^2} - (TW + 1)\sqrt{1 + 4Q^2}]}{2(TW + 1)Q} \quad (6\text{-}20)$$

and

$$\frac{TW_U}{\omega_0} = \frac{TW - [\sqrt{1 + 4(TW + 1)^2 Q^2} - (TW + 1)\sqrt{1 + 4Q^2}]}{2(TW + 1)Q} \quad (6\text{-}21)$$

where TW is the normalized low-pass transition width of the corresponding low-pass filter. For high-Q circuits a good approximation to Eqs. (6-20) and (6-21) is

$$\frac{TW_L}{\omega_0} = \frac{TW_U}{\omega_0} = \frac{TW}{2(TW + 1)Q} \quad (6\text{-}22)$$

which is the average of the two transition widths. The values of TW are given, in the case of Butterworth, Chebyshev, and inverse Chebyshev filters, by Eqs. (2-16), (2-17), and (3-7), with $\omega_c = 1$, and for elliptic filters, by Appendix C.

For a given band-reject function we may use Eqs. (6-20) and (6-21) to find the two transition widths. On the other hand, if we want the filter of lowest order with transition widths less than some prescribed amount, we may find the maximum allowable TW in the corresponding low-pass case and use this value to select the appropriate low-pass filter. The band-reject data are then found from the low-pass data. Since we may show from Eqs. (6-20) and (6-21) that $TW_U > TW_L$, we may take a value of TW_U as the maximum allowable value (since TW_L will also be less than this allowable value). From Eq. (6-21), the corresponding normalized low-pass TW is then

$$TW = \frac{2\{[2Q(TW_U)/\omega_0] - 1 - \sqrt{1 + 4Q^2}\}}{4Q^2 - \{[2Q(TW_U)/\omega_0] - 1 - \sqrt{1 + 4Q^2}\}^2} - 1 \quad (6\text{-}23)$$

SEC. 6-4   INFINITE-GAIN MFB BAND-REJECT FILTER   **137**

The use of this formula is very similar to that of Eq. (5-21) in Section 5-3, where an example was given.

6-4 INFINITE-GAIN MULTIPLE-FEEDBACK BAND-REJECT FILTER

The infinite-gain multiple-feedback (MFB) circuit [14] is one of the simpler second-order band-reject filters (Fig. 6-4). Analysis

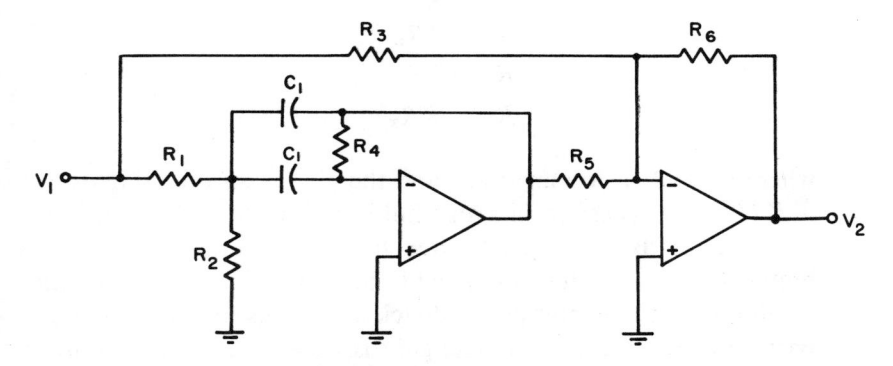

FIGURE 6-4 Infinite-gain MFB band-reject filter.

shows that it achieves Eq. (6-17) with

$$\rho = -\frac{R_6}{R_3}$$

$$\alpha\omega_0^2 = \gamma\omega_0^2 = \frac{1}{R_4 C_1^2}\left(\frac{1}{R_1} + \frac{1}{R_2}\right) \qquad (6\text{-}24)$$

$$\beta\omega_0 = \frac{2}{R_4 C_1}$$

provided that

$$2R_1 R_5 = R_3 R_4 \qquad (6\text{-}25)$$

From the second of Eqs. (6-24) we see that $\alpha = \gamma$. Thus the circuit is restricted to second-order band-reject filters with transfer function (6-7), or second-order stages with transfer function (6-6), arising from a first-order low-pass stage. In Eq. (6-6) we have

$$\rho = K, \quad \alpha = \gamma = 1, \quad \beta = \frac{1}{CQ} \qquad (6\text{-}26)$$

in which case a solution of Eqs. (6-24) and (6-25) is

$$R_1 = \frac{CQ}{2\omega_0 C_1}$$
$$R_2 = \frac{R_1}{C^2 Q^2 - 1}$$
$$R_4 = 4R_1 \qquad (6\text{-}27)$$
$$R_5 = 2R_3$$
$$R_6 = KR_3$$

where C_1 and R_3 are arbitrary and the gain is an inverting one of $-K$ ($K > 0$). These results also hold for Eq. (6-7) if $C = 1$.

The MFB band-reject filter has fewer elements than the biquad circuit, which is discussed in Section 3-5, and has the other advantages of the multiple-feedback structures in low-pass, high-pass, and bandpass cases. The gain is inverting, with magnitude R_6/R_3, and the circuit is capable of attaining Q's up to 25 [17].

The design procedure is summarized in Section 6-8.

6-5 VCVS BAND-REJECT FILTERS

The VCVS circuit [15] of Fig. 6-5 realizes the second-order band-reject function (6-17), where

$$\rho = 1$$
$$\alpha \omega_0^2 = \gamma \omega_0^2 = \frac{1}{R_1 R_2 C_1^2} \qquad (6\text{-}28)$$
$$\beta \omega_0 = \frac{2}{R_2 C_1}$$

VCVS BAND-REJECT FILTERS

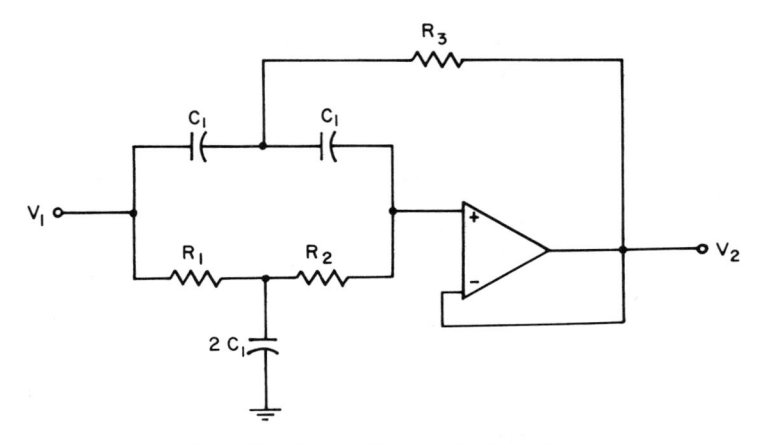

FIGURE 6-5 VCVS band-reject filter.

provided that

$$\frac{1}{R_3} = \frac{1}{R_1} + \frac{1}{R_2} \tag{6-29}$$

From the first two of Eqs. (6-28) we see that the gain of the filter is 1 and that $\alpha = \gamma$. Thus this circuit, like the MFB circuit of Section 6-4, is restricted to second-order filters or stages arising from a first-order low-pass stage.

Solving for the resistances and replacing ρ, α, β, and γ by their values in Eq. (6-6), we have

$$\begin{aligned} R_1 &= \frac{\beta}{2\alpha\omega_0 C_1} = \frac{1}{2CQ\omega_0 C_1} \\ R_2 &= \frac{2}{\beta\omega_0 C_1} = \frac{2CQ}{\omega_0 C_1} \\ R_3 &= 1\Big/\Big(\frac{1}{R_1} + \frac{1}{R_2}\Big) = \frac{R_1 R_2}{R_1 + R_2} \end{aligned} \tag{6-30}$$

where C_1 is arbitrary. The case of Eq. (6-7) is obtained by taking $C = 1$. We may select C_1 (preferably near $10/f_0$ μF) and then determine the resistance values.

Some advantages of the VCVS circuit are its requirement of a minimal number of elements and its noninverting gain. If $Q > 10$, an undesirable spread of element values results. Thus Q should be restricted to values of 10 or less for best performance. A disadvantage of the circuit is the restriction that the gain is 1.

6-6 TUNING THE SECOND-ORDER BAND-REJECT STAGES

The tuning of a second-order band-reject stage with transfer function

$$\frac{V_2}{V_1} = \frac{p(s^2 + \alpha\omega_0^2)}{s^2 + \beta\omega_0 s + \gamma\omega_0^2} \qquad (6\text{-}31)$$

is more easily facilitated if one observes the corresponding amplitude response. A second-order filter or a second-order stage arising from a first-order low-pass stage has an amplitude response like that of Fig. 6-1, except that the notch is much sharper. This is true since in this case $\alpha = \gamma = 1$.

There will be two band-reject stages like Eq. (6-31) for each second-order low-pass stage. These functions are given by Eqs. (6-9) and (6-10) for Butterworth and Chebyshev filters and by Eqs. (6-14) and (6-15) for elliptic and inverse Chebyshev filters. One of these two stages will be a low-pass stage ($\alpha > \gamma$) with an amplitude response like Fig. 3-11(a) or (b). The other is a high-pass stage ($\alpha < \gamma$) with amplitude response like Fig. 4-7(a) or (b).

In Figs. 3-11(a) and 4-7(a) the peak amplitude is

$$K_m = \frac{2p}{\beta}\sqrt{\frac{(\alpha - \gamma)^2 + \alpha\beta^2}{4\gamma - \beta^2}} \qquad (6\text{-}32)$$

which occurs at the frequency

$$f_m = f_0\sqrt{\frac{2\gamma(\alpha - \gamma) - \alpha\beta^2}{2(\alpha - \gamma) + \beta^2}} \qquad (6\text{-}33)$$

The dc amplitude is $|p|\alpha/\gamma$ and the zero magnitude occurs at $f_z = f_0\sqrt{\alpha}$ in all four cases of Figs. 3-11 and 4-7.

Tuning procedures for the individual band-reject circuits will be given in the design summaries to follow.

6-7 GENERAL DESIGN INFORMATION FOR BAND-REJECT FILTER CONSTRUCTION

There are basically four types of second-order band-reject filter stages that one may encounter in constructing a band-reject filter. Each of these types has a transfer function like Eq. (6-17), which we repeat with $p = K$ as

$$\frac{V_2}{V_1} = \frac{K(s^2 + \alpha\omega_0^2)}{s^2 + \beta\omega_0 s + \gamma\omega_0^2}$$

The parameters α, β, and γ are different for each of the four types.

Type I is that of the second-order band-reject filter consisting of a single stage with $\alpha = \gamma = 1$ and $\beta = 1/Q$.

Type II is a stage corresponding to a first-order low-pass stage, for which $\alpha = \gamma = 1$ and $\beta = 1/CQ$, where C is the normalized coefficient in the corresponding first-order low-pass function given in Appendix A, B, or C. This type is a stage in higher-order band-reject filters of Butterworth, Chebyshev, inverse Chebyshev, or elliptic types.

Type III arises from second-order low-pass stages of Butterworth or Chebyshev filters. For each second-order low-pass stage there are two band-reject stages. The first has $\alpha = 1$, $\beta = D_1/E_1$, and $\gamma = D_1^2$, and the second has $\alpha = 1$, $\beta = 1/D_1 E_1$, and $\gamma = 1/D_1^2$, where E_1 and D_1, given in Eqs. (6-11) and (6-12), depend on the low-pass coefficients B and C, found in Appendix A.

Type IV arises from second-order low-pass stages of inverse Chebyshev or elliptic filters. As in type III, there are two band-reject stages for each low-pass stage. The first has $\alpha = A_2$, $\beta = D_1/E_1$, and $\gamma = D_1^2$, and the second has $\alpha = 1/A_2$, $\beta = 1/D_1 E_1$, and $\gamma = 1/D_1^2$. where E_1, D_1, and A_2, given in Eqs. (6-11), (6-12), and (6-16), depend on the low-pass coefficients A, B, and C, found in Appendix B or C.

As in the case of bandpass filters, band-reject filters are constructed by designing and cascading second-order stages. For example, a sixth-order elliptic band-reject filter corresponds to a third-order elliptic low-pass filter, which has one first-order section and one second-order section. Thus the band-reject filter will consist of one stage of type II, corresponding to the first-order low-pass stage, cascaded with two stages of type IV, corresponding to the second-order low-pass stage.

The MFB and VCVS second-order band-reject filters of Sections 6-4 and 6-5 can be used to realize only the transfer functions of types I and II. This is because these circuits are constrained to the case $\alpha = \gamma = 1$. However, the elliptic low-pass functions of Chapter 3 have the same form as the band-reject functions of types III and IV (as well as I and II). Thus the low-pass elliptic circuits of Chapter 3 may be used to realize any of the four band-reject types. These circuits, the VCVS elliptic circuit of Section 3-3, the three-capacitor circuit of Section 3-4, and the biquad elliptic circuit of Section 3-5, will be discussed along with those of Sections 6-4 and 6-5, in the design summaries at the end of the chapter.

6-8 MFB BAND-REJECT FILTER DESIGN SUMMARY

To design (a) a second-order band-reject filter, or (b) a second-order stage of a higher-order band-reject filter corresponding to a first-order low-pass stage, having a given center frequency f_0 Hz (or $\omega_0 = 2\pi f_0$ rad/s), stage gain K, and quality factor Q, perform the following steps.

1. In case (a) take $C = 1$, and in case (b) find the normalized coefficient C of the first-order low-pass stage from the appropriate table of Appendix A, B, or C

2. Select a standard value of C_1 (preferably near $10/f_0$ μF) and calculate the resistance values given by

$$R_1 = \frac{CQ}{2\omega_0 C_1}$$

$$R_2 = \frac{R_1}{C^2 Q^2 - 1}$$

$$R_4 = 4R_1$$

$$R_5 = 2R_3$$

$$R_6 = KR_3$$

SEC. 6-8   MFB BAND-REJECT FILTER DESIGN   143

The value of R_3 is arbitrary and may be chosen to minimize the resistance spread. For moderate gain a reasonable value is $R_3 = 1/\omega_0 C_1$

3. Select standard values of the resistances as close as possible to the calculated values and construct the filter, or its stages, in accordance with Fig. 6-6. (See also Section 6-7 for general information on the various filter stages.)

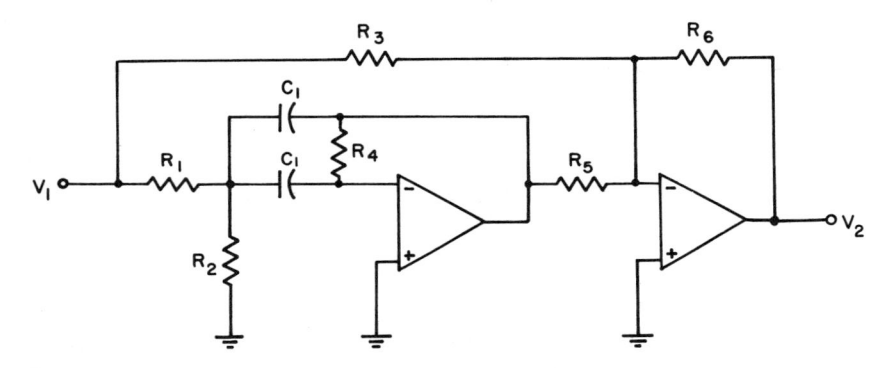

FIGURE 6-6   MFB band-reject filter circuit.

Comments

(a) For best performance, element values close to those selected and calculated should be used. Higher-order filters require more accurate element values than lower-order filters. The performance of the filter is unchanged if all the resistances are multiplied and the capacitances divided by a common factor

(b) The input impedance of each op amp should be at least $10 R_{eq}$, where R_{eq} is R_4 or R_5 connected to its inverting input terminal. The open-loop gain of each op amp should be at least 50 times the amplitude of the filter, or stage, at f_a, the highest desired frequency in the passband, and its slew rate (volts per microsecond) should be at least $\frac{1}{2}\omega_a \times 10^{-6}$ times the peak-to-peak output voltage

(c) The gain is inverting with magnitude

$$K = \frac{R_6}{R_3}$$

Thus the gain can be adjusted by varying R_6. The quality factor Q may be adjusted, without affecting f_0, by varying R_4 (see also Section 6-6)

(d) The circuit may be used for values of Q up to 25

(e) The order required for given upper and lower transition widths or, conversely, the transition widths ensuing from a given order, may be found from Section 6-3

The MFB band-reject filter was discussed in Section 6-4.

6-9 VCVS BAND-REJECT FILTER DESIGN SUMMARY

To design (a) a second-order band-reject filter, or (b) a second-order stage of a higher-order band-reject filter corresponding to a first-order low-pass stage, having a given center frequency f_0 Hz (or $\omega_0 = 2\pi f_0$ rad/s), stage gain $K = 1$, and quality factor Q, perform the following steps.

1. In case (a) take $C = 1$, and in case (b) find the normalized coefficient C of the first-order low-pass stage from the appropriate table of Appendix A, B, or C

2. Select a standard value of C_1 (preferably near $10/f_0$ µF) and calculate the resistance values given by

$$R_1 = \frac{1}{2\omega_0 CQC_1}$$

$$R_2 = \frac{2CQ}{\omega_0 C_1}$$

$$R_3 = \frac{R_1 R_2}{R_1 + R_2}$$

3. Select standard values of the resistances as close as possible to the calculated values and construct the filter, or its stages, in accordance with Fig. 6-7. (See also Section 6-7 for general information on the various filter stages.)

SEC. 6-9   VCVS BAND-REJECT FILTER DESIGN   145

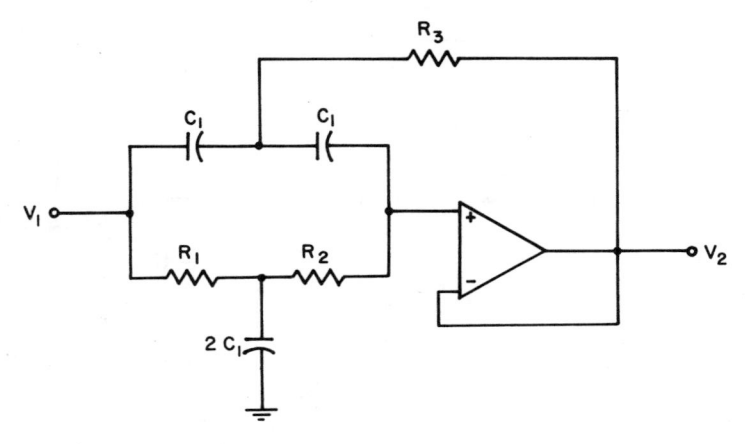

FIGURE 6-7 VCVS band-reject filter.

Comments

(a) Comments (a) and (b) for the MFB filter of Section 6-8 apply directly, except that in (b), $R_{eq} = R_1 + R_2$

(b) There must be a dc return to ground at the filter input

(c) The center frequency f_0 can be adjusted without affecting Q, by varying R_1 (see also Section 6-6)

(d) The circuit should be restricted to values of Q of 10 or less

The VCVS band-reject filter was discussed in Section 6-5.

6-10 VCVS BAND-REJECT ELLIPTIC FILTER DESIGN SUMMARY

The VCVS band-reject elliptic filter circuit of Fig. 6-8 can be used to design (a) a second-order band-reject filter, (b) a second-order band-reject stage of a higher-order filter corresponding to a first-order stage of a low-pass Butterworth, Chebyshev, inverse Chebyshev, or elliptic filter, (c) the two second-order band-reject stages of a higher-order filter corresponding to a second-order

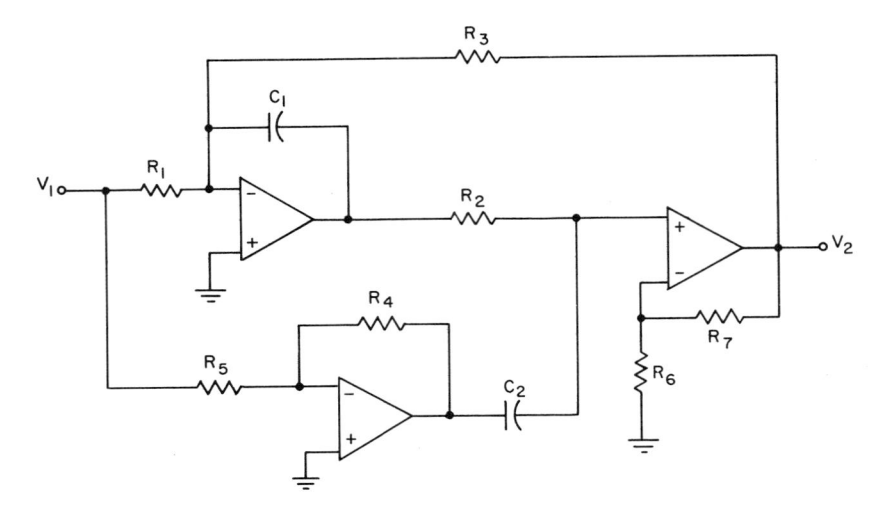

FIGURE 6-8 VCVS band-reject elliptic filter circuit.

stage of a low-pass Butterworth or Chebyshev filter, or (d) the two second-order band-reject stages of a higher-order filter corresponding to a second-order stage of a low-pass inverse Chebyshev or elliptic filter. To design the filter, or its stages, having a given center frequency f_0 Hz (or $\omega_0 = 2\pi f_0$ rad/s), gain K, quality factor Q, minimum stopband loss (MSL) in the case of the inverse Chebyshev and elliptic filters, and passband ripple width (PRW) in the case of the elliptic filter, perform the following steps.

1. In case (b) find the normalized coefficient C of the first-order low-pass stage from the appropriate table of Appendix A, B, or C. In case (c) find the normalized low-pass coefficients B and C from the appropriate table of Appendix A. In case (d) find the normalized low-pass coefficients A, B, and C from the appropriate table of Appendix B or C

2. Select a standard value of C_1 (preferably near $10/f_0$ μF) and calculate the resistance values given by

$$R_1 = \frac{\mu\beta}{K\alpha\omega_0 C_1}$$

$$R_2 = \frac{1}{\beta\omega_0 C_2}$$

SEC. 6-10 VCVS BAND-REJECT FILTER DESIGN 147

$$R_3 = \frac{K\alpha R_1}{\gamma}$$

$$R_4 = \frac{KR_5}{\mu}$$

$$R_6 = \frac{\mu R_2}{\mu - 1}$$

$$R_7 = \mu R_2$$

where C_2, R_5, and $\mu > 1$ are arbitrary. In case (a)

$$\alpha = \gamma = 1$$

$$\beta = \frac{1}{Q}$$

and in case (b)

$$\alpha = \gamma = 1$$

$$\beta = \frac{1}{CQ}$$

In case (c) there are two stages. The first stage has

$$\alpha = 1$$

$$\beta = \frac{D_1}{E_1}$$

$$\gamma = D_1^2$$

and the second has

$$\alpha = 1$$

$$\beta = \frac{1}{D_1 E_1}$$

$$\gamma = \frac{1}{D_1^2}$$

Also in case (d) there are two stages. The first stage has

$$\alpha = A_2$$

$$\beta = \frac{D_1}{E_1}$$

$$\gamma = D_1^2$$

and the second has

$$\alpha = \frac{1}{A_2}$$

$$\beta = \frac{1}{D_1 E_1}$$

$$\gamma = \frac{1}{D_1^2}$$

where

$$A_2 = 1 + \frac{1}{2AQ^2}(1 + \sqrt{1 + 4AQ^2})$$

$$E_1 = \frac{1}{B}\sqrt{\frac{C}{2}[1 + 4CQ^2 + \sqrt{(1 + 4CQ^2)^2 - (2BQ)^2}]}$$

$$D_1 = \frac{1}{2}\left[\frac{BE_1}{QC} + \sqrt{\left(\frac{BE_1}{QC}\right)^2 - 4}\right]$$

If K and the quality factor [Q in cases (a) and (b) and E_1 in cases (c) and (d)] are moderate values such as 10 or less, reasonable values of the arbitrary elements are

$$C_2 = C_1$$

$$R_5 = \frac{1}{\omega_0 C_1}$$

and $\mu = 2$ (in which case $R_6 = R_7$). If the quality factor and/or K are high, say over 10, then C_2, R_5, and μ should be chosen to maintain a lower spread of resistance values

3. Select standard values of resistances as close as possible to the calculated values and construct the filter, or its stages, in accordance with Fig. 6-8

4. If $\mu = 1$ is desired, then R_6 becomes an open circuit and R_7 becomes a short circuit, with the other resistances given as in step 2. In this case the circuit is the voltage follower circuit of Fig. 3-6

SEC. 6-11 3-CAPACITOR BR ELLIPTIC FILTER DESIGN

Comments

(a) Comments (a), (b), and (e) for the MFB filter of Section 6-8 apply directly, except that in (b), R_{eq} for each op amp is the resistance R_1, R_2, or R_5 connected to its input terminal

(b) The circuit may be used for both high and low pole-pair Q (or E_1) by selecting the arbitrary parameters of step 2 to keep the spread of resistance values relatively low. For moderate gains, Q (or E_1) may range as high as 100

(c) Tuning may be accomplished by varying the ratio R_4/R_5 to move the notch toward f_z, shown in Fig. 3-11. The gain μ of the VCVS, given by

$$\mu = 1 + \frac{R_7}{R_6}$$

may then be adjusted, by varying the ratio R_7/R_6, to move the peak toward f_m. Finally, R_2 may be varied to obtain K_m. These steps may be repeated until the stage is tuned

(d) The circuit yields an inverting gain $-K$ ($K > 0$)

The VCVS elliptic filter was discussed in Section 3-3.

6-11 THREE-CAPACITOR BAND-REJECT ELLIPTIC FILTER DESIGN SUMMARY

The three-capacitor band-reject elliptic filter circuit of Fig. 6-9 can be used to design (a) a second-order band-reject filter, (b) a second-order band-reject stage of a higher-order filter corresponding to a first-order stage of a low-pass Butterworth, Chebyshev, inverse Chebyshev, or elliptic filter, (c) the two second-order band-reject stages of a higher-order filter corresponding to a second-order stage of a low-pass Butterworth or Chebyshev filter, or (d) the two second-order band-reject stages of a higher-order filter corresponding to a second-order stage of a low-pass inverse Chebyshev or elliptic filter. To design the filter, or its stages, having a given center frequency f_0 Hz (or $\omega_0 = 2\pi f_0$ rad/s), gain K, quality

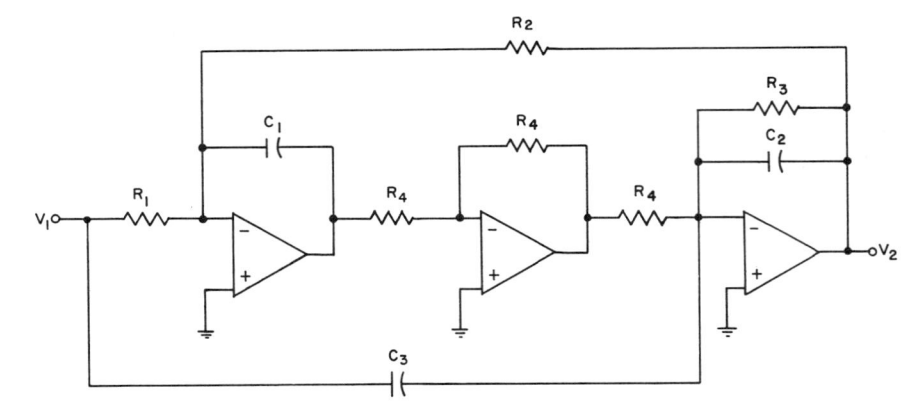

FIGURE 6-9 *Three-capacitor band-reject elliptic filter circuit.*

factor Q, minimum stopband loss (MSL) in the case of the inverse Chebyshev and elliptic filters, and passband ripple width (PRW) in the case of the elliptic filter, perform the following steps.

1. In case (b) find the normalized coefficient C of the first-order low-pass stage from the appropriate table of Appendix A, B, or C. In case (c) find the normalized low-pass coefficients B and C from the appropriate table of Appendix A. In case (d) find the normalized low-pass coefficients A, B, and C from the appropriate table of Appendix B or C

2. Select a standard value of C_1 (preferably near $10/f_0$ μF) and calculate the element values given by

$$C_3 = KC_2$$

$$R_1 = \frac{1}{R_4 K\alpha\omega_0^2 C_1 C_2}$$

$$R_2 = \frac{K\alpha R_1}{\gamma}$$

$$R_3 = \frac{1}{\beta\omega_0 C_2}$$

where C_2 and R_4 are arbitrary. In case (a)

$$\alpha = \gamma = 1$$

$$\beta = \frac{1}{Q}$$

SEC. 6-11   3-CAPACITOR BR ELLIPTIC FILTER DESIGN

and in case (b)

$$\alpha = \gamma = 1$$
$$\beta = \frac{1}{CQ}$$

In case (c) there are two stages. The first stage has

$$\alpha = 1$$
$$\beta = \frac{D_1}{E_1}$$
$$\gamma = D_1^2$$

and the second has

$$\alpha = 1$$
$$\beta = \frac{1}{D_1 E_1}$$
$$\gamma = \frac{1}{D_1^2}$$

Also in case (d) there are two stages. The first stage has

$$\alpha = A_2$$
$$\beta = \frac{D_1}{E_1}$$
$$\gamma = D_1^2$$

and the second has

$$\alpha = \frac{1}{A_2}$$
$$\beta = \frac{1}{D_1 E_1}$$
$$\gamma = \frac{1}{D_1^2}$$

where

$$A_2 = 1 + \frac{1}{2AQ^2}(1 + \sqrt{1 + 4AQ^2})$$

$$E_1 = \frac{1}{B}\sqrt{\frac{C}{2}[1 + 4CQ^2 + \sqrt{(1 + 4CQ^2)^2 - (2BQ)^2}]}$$

$$D_1 = \frac{1}{2}\left[\frac{BE_1}{QC} + \sqrt{\left(\frac{BE_1}{QC}\right)^2 - 4}\right]$$

If K and the quality factor [Q in cases (a) and (b) and E_1 in cases (c) and (d)] are moderate values such as 10 or less, reasonable values of the arbitrary elements are

$$C_2 = C_1$$

$$R_4 = \frac{1}{\omega_0 C_1}$$

If the quality factor and/or K are high, say over 10, then C_2 and R_4 should be chosen to maintain a lower spread of resistance values

3. Select standard values of resistances and capacitances as close as possible to the calculated values and construct the filter, or its stages in accordance with Fig. 6-9

Comments

(a) Comments (a), (b), and (e) for the MFB filter of Section 6-8 apply directly, except that in (b), R_{eq} for each op amp is the resistance R_1 or R_4 connected to its inverting input

(b) The circuit may be used for both low and high pole-pair Q (or E_1), with an upper limit of approximately $Q = 100$

(c) Tuning may be accomplished as discussed in Section 6-6 by first adjusting R_1 until the notch is at f_z, then adjusting R_2 to place the peak amplitude at f_m, and finally adjusting R_3 to obtain the correct value of K_m. For low-Q sections the last two steps may need to be repeated

(d) The circuit yields an inverting gain $-K$ ($K > 0$)

The three-capacitor circuit was discussed in Section 3-4.

6-12 BIQUAD BAND-REJECT ELLIPTIC FILTER DESIGN SUMMARY

The biquad band-reject elliptic filter circuit of Fig. 6-10 can be used to design (a) a second-order band-reject filter, (b) a second-order band-reject stage of a higher-order filter corresponding to a first-order stage of a low-pass Butterworth, Chebyshev, inverse Chebyshev, or elliptic filter, (c) the two second-order band-reject stages of a higher-order filter corresponding to a second-order stage of a low-pass Butterworth or Chebyshev filter, or (d) the two second-order band-reject stages of a higher-order filter corresponding to a second-order stage of a low-pass inverse Chebyshev or elliptic filter. To design the filter, or its stages, having a given center frequency f_0 Hz (or $\omega_0 = 2\pi f_0$ rad/s), gain K, quality factor Q, minimum stopband loss (MSL) in the case of the inverse Chebyshev and elliptic filters, and passband ripple width (PRW) in the case of the elliptic filter, perform the following steps.

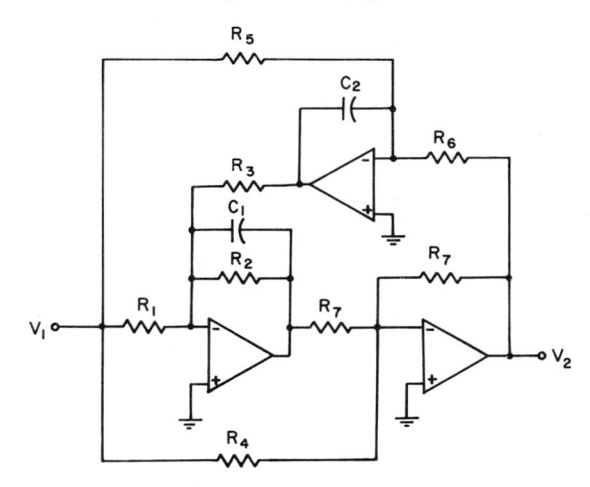

FIGURE 6-10 Biquad elliptic low-pass filter circuit.

1. In case (b) find the normalized coefficient C of the first-order low-pass stage from the appropriate table of Appendix A, B, or C. In case (c) find the normalized low-pass coefficients B and C from the appropriate table of Appendix A. In case (d) find the normalized low-pass coefficients A, B, and C from the appropriate table of Appendix B or C

2. Select a standard value of C_1 (preferably near $10/f_0$ μF) and calculate the resistance values given by

$$R_1 = \frac{1}{K\beta\omega_0 C_1}$$

$$R_2 = KR_1$$

$$R_3 = \frac{1}{\sqrt{\gamma}\,\omega_0 C_1}$$

$$R_4 = \frac{R_7}{K}$$

$$R_5 = \frac{\sqrt{\gamma}}{K\alpha\omega_0 C_2}$$

$$R_6 = \frac{C_1 R_3}{C_2}$$

where C_2 and R_7 are arbitrary. In case (a)

$$\alpha = \gamma = 1$$

$$\beta = \frac{1}{Q}$$

and in case (b)

$$\alpha = \gamma = 1$$

$$\beta = \frac{1}{CQ}$$

In case (c) there are two stages. The first stage has

$$\alpha = 1$$

$$\beta = \frac{D_1}{E_1}$$

$$\gamma = D_1^2$$

and the second has

$$\alpha = 1$$

$$\beta = \frac{1}{D_1 E_1}$$

$$\gamma = \frac{1}{D_1^2}$$

SEC. 6-12   BIQUAD BR ELLIPTIC FILTER DESIGN   155

Also in case (d) there are two stages. The first stage has

$$\alpha = A_2$$
$$\beta = \frac{D_1}{E_1}$$
$$\gamma = D_1^2$$

and the second has

$$\alpha = \frac{1}{A_2}$$
$$\beta = \frac{1}{D_1 E_1}$$
$$\gamma = \frac{1}{D_1^2}$$

where

$$A_2 = 1 + \frac{1}{2AQ^2}(1 + \sqrt{1 + 4AQ^2})$$

$$E_1 = \frac{1}{B}\sqrt{\frac{C}{2}[1 + 4CQ^2 + \sqrt{(1 + 4CQ^2)^2 - (2BQ)^2]}}$$

$$D_1 = \frac{1}{2}\left[\frac{BE_1}{QC} + \sqrt{\left(\frac{BE_1}{QC}\right)^2 - 4}\right]$$

If K and the quality factor [Q in cases (a) and (b) and E_1 in cases (c) and (d)] are moderate values such as 10 or less, reasonable values of the arbitrary elements are

$$C_2 = C_1$$
$$R_7 = \frac{1}{\omega_0 C_1}$$

If the quality factor and/or K are high, say over 10, then C_2 and R_7 should be chosen to maintain a lower spread of resistance values

3. Select standard values of resistances and capacitances as close as possible to the calculated values and construct the filter, or its stages in accordance with Fig. 6-10

Comments

(a) Comments (a), (b), and (e) for the MFB filter of Section 6-8 apply directly, except that in (b), R_{eq} for each op amp is the resistance R_1, R_6, or R_7 connected to its input terminal

(b) The circuit may be used for both high and low pole-pair Q (or E_1) by selecting the arbitrary parameters of step 2 to keep the spread of resistance values relatively low. For moderate gains, Q (or E_1) may range as high as 100

(c) Tuning may be accomplished by adjusting R_4 to set the notch at f_z, adjusting R_3 to set the center frequency, adjusting R_2 to set Q, and adjusting R_1 or R_5 to set the gain (see also Section 6-6)

(d) The circuit yields an inverting gain with magnitude R_2/R_1

The biquad elliptic filter was discussed in Section 3-5.

7

ALL-PASS AND CONSTANT-TIME-DELAY FILTERS

7-1 ALL-PASS FILTERS

Up to now we have been interested in frequency-selective filters, for which the amplitude response is the important factor. In this final chapter we shall consider other types of filters for which the phase response and/or its associated time delay are the factors of interest.

The first filter we shall consider is the *all-pass*, or *phase-shifting*, filter, which has a constant amplitude $|H(j\omega)| = K$ and a phase $\phi(\omega)$ that is a function of frequency. A typical phase response is shown in Fig. 7-1, where it may be seen that if ϕ_0 is the phase, or phase shift, at ω_0 rad/s or $f_0 = \omega_0/2\pi$ Hz, then

$$\phi(\omega_0) = \phi_0$$

Since the transfer function we are considering is $H = V_2/V_1$, the phase shift ϕ_0 at $\omega = \omega_0$ is the difference between the phase of

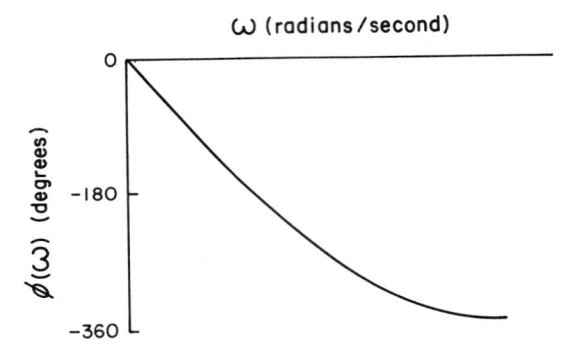

FIGURE 7-1 *Typical phase response.*

V_2 and the phase of V_1. That is, the phase of the output V_2 is greater than that of the input V_1 by ϕ_0 degrees. If the two waveforms are sinusoids, the output wave reaches its peaks or dips ϕ_0 degrees, or if ϕ_0 is expressed in radians, ϕ_0/ω_0 seconds, before the input wave reaches its peaks or dips. Therefore, the output wave *leads* the input wave (or the input *lags* the output) by ϕ_0. (In most cases, however, ϕ_0 is negative, so that the output actually lags the input by a positive angle.)

The difference in seconds ϕ_0/ω_0 between the two successive input and output peaks is approximated by the *time delay*, which we defined in Section 1-2. This characteristic is of importance in time-delay filters, such as the *Bessel filter*, where the emphasis is on obtaining time delays that are very nearly constant. These filters will be discussed in the latter part of the chapter.

We shall consider only second-order all-pass filters, for which the transfer function is of the form

$$H = \frac{V_2}{V_1} = \frac{K(s^2 - a\omega_0 s + b\omega_0^2)}{s^2 + a\omega_0 s + b\omega_0^2} \tag{7-1}$$

where K, a, and b are suitably chosen constants, and ω_0 is the frequency of interest. In the case of Eq. (7-1), we have $|H(j\omega)| = K$, as required, and the phase response given by

$$\phi(\omega) = -2 \arctan\left(\frac{a\omega_0 \omega}{b\omega_0^2 - \omega^2}\right) \tag{7-2}$$

The constant K, which is the *gain* of the filter, and the coefficients a and b, which determine the phase shift, are specified by the filter requirements. As in the other filter types, there is also a pole-pair Q, defined by $Q = \sqrt{b}/a$.

Evidently, if $\phi(\omega_0) = \phi_0$ is the specified phase shift at ω_0, we have, by Eq. (7-2),

$$\phi_0 = -2 \arctan\left(\frac{a}{b-1}\right) \qquad (7\text{-}3)$$

and therefore we have two parameters, a and b, to be used to determine a given condition ϕ_0. Thus we have an extra parameter which may be used to determine an additional condition, such as setting the gain of the filter, or minimizing the dc offset of an op amp.

7-2 MULTIPLE-FEEDBACK ALL-PASS FILTERS

A circuit that realizes the second-order all-pass function of Eq. (7-1) is that of Fig. 7-2 [7], which we shall refer to as a multiple-feedback (MFB) circuit. The all-pass function (7-1) is achieved with

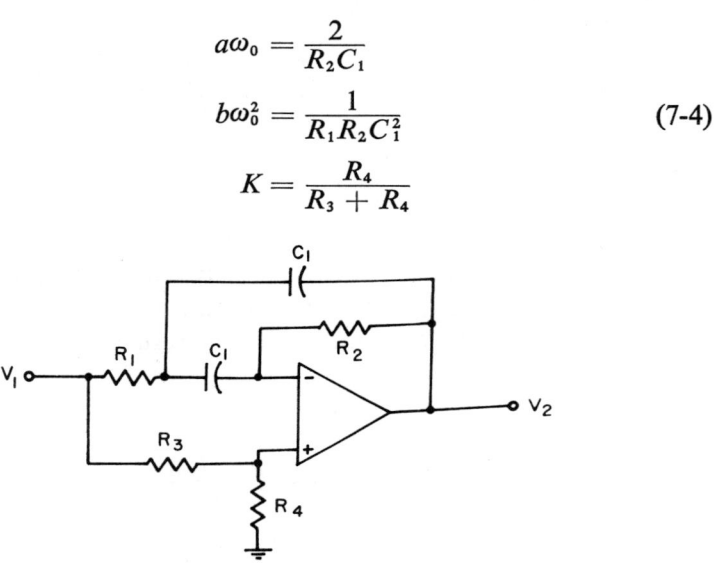

$$a\omega_0 = \frac{2}{R_2 C_1}$$

$$b\omega_0^2 = \frac{1}{R_1 R_2 C_1^2} \qquad (7\text{-}4)$$

$$K = \frac{R_4}{R_3 + R_4}$$

FIGURE 7-2 Multiple feedback all-pass filter.

provided that we have

$$4R_1R_4 = R_2R_3 \tag{7-5}$$

Evidently, the gain K is restricted to

$$0 < K < 1 \tag{7-6}$$

In addition, if we wish to minimize the dc offset of the op amp, we have

$$\frac{R_3R_4}{R_3 + R_4} = R_2 \tag{7-7}$$

From Eqs. (7-4) and (7-5) we obtain

$$K = \frac{b}{b + a^2} \tag{7-8}$$

Thus, for a given $\phi_0 = \phi(\omega_0)$ and K, we have, by Eqs. (7-3) and (7-8), for $0 < \phi_0 < 180°$,

$$a = \frac{(1-K)\{-1 + \sqrt{1 + [4K/(1-K)]\tan^2(\phi_0/2)}\}}{2K \tan(\phi_0/2)} \tag{7-9}$$

and for $-180° < \phi_0 < 0$,

$$a = \frac{(1-K)\{-1 - \sqrt{1 + [4K/(1-K)]\tan^2(\phi_0/2)}\}}{2K \tan(\phi_0/2)} \tag{7-10}$$

and

$$b = \frac{a^2K}{1-K} \tag{7-11}$$

To construct the circuit we may choose C_1 (preferably near $10/f_0$ μF, where $f_0 = \omega_0/2\pi$) and calculate the resistance values from Eqs. (7-4), (7-5), and (7-7). The results are

$$\begin{aligned}R_2 &= \frac{2}{a\omega_0 C_1} \\ R_1 &= \frac{1-K}{4K}R_2 \\ R_3 &= \frac{R_2}{K} \\ R_4 &= \frac{R_2}{1-K}\end{aligned} \tag{7-12}$$

SEC. 7-3 BIQUAD ALL-PASS FILTERS 161

For the special case $K = \frac{1}{2}$, these values reduce to

$$R_2 = \frac{2}{a\omega_0 C_1}$$

$$R_1 = \frac{R_2}{4} \qquad (7\text{-}13)$$

$$R_3 = R_4 = 2R_2$$

In this case Eqs. (7-9) and (7-10) become, for $0 < \phi_0 < 180°$,

$$a = \frac{-1 + \sqrt{1 + 4\tan^2(\phi_0/2)}}{2\tan(\phi_0/2)} \qquad (7\text{-}14)$$

and for $-180° < \phi_0 < 0$,

$$a = \frac{-1 - \sqrt{1 + 4\tan^2(\phi_0/2)}}{2\tan(\phi_0/2)} \qquad (7\text{-}15)$$

As an example, suppose that we want an all-pass filter to provide a phase shift of $\phi_0 = -90°$ at $f_0 = 1000$ Hz, with a gain of $K = \frac{1}{2}$. From Eq. (7-15) we have

$$a = \frac{-1 - \sqrt{5}}{-2} = 1.618$$

Choosing $C_1 = 10/1000 = 0.01$ μF, we have, by Eq. (7-13),

$$R_2 = 19.673 \text{ k}\Omega$$
$$R_1 = 4.918 \text{ k}\Omega$$
$$R_3 = R_4 = 39.346 \text{ k}\Omega$$

The MFB all-pass circuit is summarized and a design procedure given in Section 7-6.

7-3 BIQUAD ALL-PASS FILTERS

The all-pass circuit of the previous section was restricted to gains less than 1. A biquad all-pass circuit [8] that can achieve an arbitrary gain and a pole-pair Q up to 100 or so, and which is relatively easy to tune, is shown in Fig. 7-3. The circuit achieves

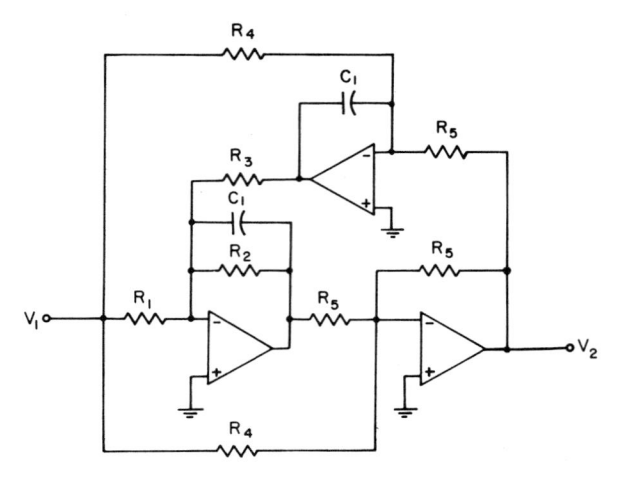

FIGURE 7-3 Biquad all-pass filter.

Eq. (7-1) with

$$a\omega_0 = \frac{1}{R_2 C_1}$$
$$b\omega_0^2 = \frac{1}{R_3 R_5 C_1^2} \tag{7-16}$$

and an inverting gain $-K$ ($K > 0$) of

$$K = \frac{R_5}{R_4} \tag{7-17}$$

provided that

$$R_2 R_4 = 2 R_1 R_5 \tag{7-18}$$

For a chosen value of C_1, Eqs. (7-16), (7-17), and (7-18) are satisfied by the resistance values,

$$R_2 = \frac{1}{a\omega_0 C_1}$$
$$R_1 = \frac{R_2}{2K}$$
$$R_3 = \frac{1}{b\omega_0^2 C_1^2 R_5} \tag{7-19}$$
$$R_4 = \frac{R_5}{K}$$

where R_5 is arbitrary.

SEC. 7-4   BESSEL FILTERS   163

For a given $\phi_0 = \phi(\omega_0)$, the parameters a and b are related by Eq. (7-3). Thus, one may be chosen arbitrarily. If we choose

$$b = a^2 \qquad (7\text{-}20)$$

then a is given by Eq. (7-14) or (7-15). In this case a considerable simplification in the resistance values may be effected by choosing $R_5 = R_2$. The resulting values are

$$\begin{aligned} R_2 &= \frac{1}{a\omega_0 C_1} \\ R_1 &= \frac{R_2}{2K} \\ R_3 &= R_5 = R_2 \\ R_4 &= \frac{R_2}{K} \end{aligned} \qquad (7\text{-}21)$$

The biquad circuit may be tuned by varying R_2 to adjust a and R_3 to adjust b, which in turn adjusts ϕ_0; then varying R_1 adjusts the gain.

The design procedure is given in Section 7-7.

7-4 BESSEL FILTERS

If the phase response of Fig. 7-1 is a straight line defined by

$$\phi(\omega) = -\omega\tau \qquad (7\text{-}22)$$

where τ is a constant, then from Eq. (1-5) we have the time delay, given by

$$T(\omega) = -\frac{d}{d\omega}\phi(\omega) = \tau \qquad (7\text{-}23)$$

Thus a linear phase response (a straight line) is accompanied by a constant time delay, which is very useful in many filter applications. A filter for which the time delay is very nearly constant (over some prescribed frequency range $0 \leq \omega \leq \omega_c$) is therefore a *linear-phase* or *constant-time-delay filter*.

The best of the all-pole constant-time-delay filters is the *Bessel filter* [27], with transfer function

$$\frac{V_2}{V_1} = \frac{Kb_0}{B_n(s)} \qquad (7\text{-}24)$$

where K is the gain and $B_n(s)$ is an nth-degree polynomial given by

$$B_n(s) = s^n + b_{n-1}s^{n-1} + \cdots + b_1 s + b_0 \qquad (7\text{-}25)$$

where for $k = 0, 1, 2, \ldots, n$,

$$b_k = \frac{(2n-k)!}{k!(n-k)!}\left(\frac{\omega_c}{2}\right)^{n-k} \qquad (7\text{-}26)$$

The polynomial $B_n(s)$, with $\omega_c = 1$, is related to the *Bessel polynomials*, from which the name of the filter is derived.

The time delay of the Bessel filter is maximally flat, like the amplitude response of the Butterworth filter. This was shown by Thomson [30], who first developed the theory but did not relate his results to the Bessel polynomials. Because of Thomson's work, another name for the filter is the *Thomson filter*.

To illustrate the linearity of the Bessel phase response, we have shown a number of cases in Fig. 7-4. These may be compared with the Butterworth and Chebyshev phase responses shown earlier in Fig. 2-9. It is evident that the Bessel responses are far superior. On the other hand, the amplitude response of the Bessel filter is inferior to those of the Butterworth or Chebyshev filters.

To further illustrate the linear phase and constant time delay properties of the Bessel filter, we may show that for

$$0 \leq \omega \leq \omega_c \qquad (7\text{-}27)$$

the time delay monotonically decreases from its value at $\omega = 0$ of

$$T(0) = \frac{1}{\omega_c} \qquad (7\text{-}28)$$

SEC. 7-4 BESSEL FILTERS 165

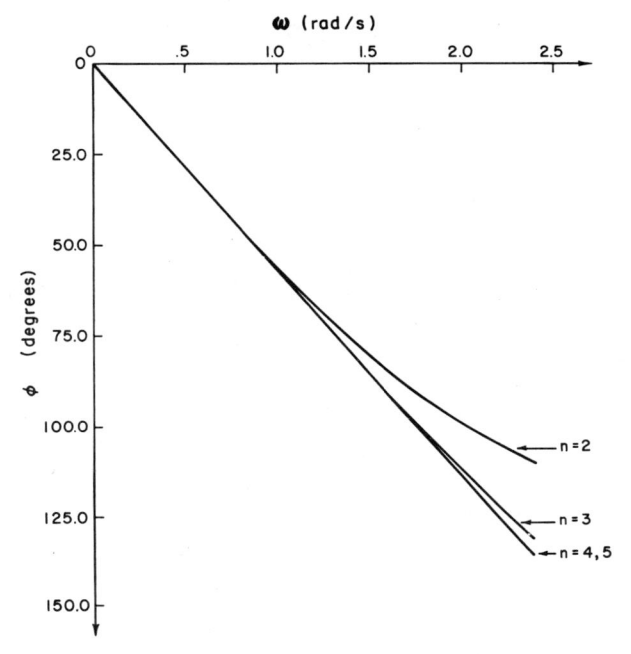

FIGURE 7-4 Bessel phase responses.

to its value at $\omega = \omega_c$, given for $n = 2$ by

$$T(\omega_c) = \frac{12}{13\omega_c} = \frac{0.92308}{\omega_c} \tag{7-29}$$

for $n = 3$ by

$$T(\omega_c) = \frac{276}{277\omega_c} = \frac{0.99639}{\omega_c} \tag{7-30}$$

and for $n = 4$ by

$$T(\omega_c) = \frac{12{,}745}{12{,}746\omega_c} = \frac{0.99992}{\omega_c} \tag{7-31}$$

and so forth. Thus as the order increases, the time delay becomes more nearly constant. The time delay is reduced by only 1% of its value $T(0)$ for $n = 5$ when $\omega = 2.71\omega_c$ and for $n = 6$ when $\omega = 3.52\omega_c$ [33].

Since the Bessel filter is an all-pole low-pass filter, it has a transfer function similar to those of the Butterworth and Chebyshev filters of Chapter 2. The transfer function may be factored into second-order functions of the form

$$\frac{V_2}{V_1} = \frac{KC\omega_c^2}{s^2 + B\omega_c s + C\omega_c^2} \tag{7-32}$$

with one first-order function (if n is odd) of the form

$$\frac{V_2}{V_1} = \frac{KC\omega_c}{s + C\omega_c} \tag{7-33}$$

The stage gain is K in each case and the coefficients B and C for the various stages are listed in Appendix E for orders $n = 2, 3, \ldots, 6$.

The amplitude response of a Bessel filter decreases monotonically from its maximum value, which occurs at zero frequency. Thus it resembles that of a Butterworth filter except that the rate of attenuation is much slower. The frequency ω_c in Eqs. (7-27) through (7-31) is not the cutoff frequency but the frequency determining the constant-time-delay band. For a given time delay $\tau = T(\omega_c)$ we may find ω_c or $f_c = \omega_c/2\pi$ approximately from

$$f_c = \frac{1}{2\pi\tau} = \frac{0.15915}{\tau} \text{ Hz} \tag{7-34}$$

The cutoff frequency, or 3-dB frequency, is given approximately, for $n \geq 3$ [33], by

$$\omega_{3\text{ dB}} = \sqrt{0.69315(2n - 1)} \tag{7-35}$$

Since the Bessel filter is an all-pole low-pass filter, it may be constructed using the methods of Chapter 2. Design procedures are summarized in Section 7-8.

7-5 ALL-PASS CONSTANT-TIME-DELAY FILTERS

The Bessel filter has a constant time delay but its amplitude response monotonically decreases with frequency. An *all-pass*, constant-time-delay filter, with a constant-time-delay frequency

SEC. 7-5 ALL-PASS CONSTANT-TIME-DELAY FILTERS

range which is twice that of the Bessel filter [16], has a transfer function given by

$$\frac{V_2}{V_1} = \frac{KB_n(-s/2)}{B_n(s/2)} \qquad (7\text{-}36)$$

where $B_n(s)$ is given by Eqs. (7-25) and (7-26). This filter has the advantage over the Bessel filter of the constant-amplitude response of an all-pass filter.

The transfer function of the all-pass, constant-time-delay filter may be factored into second-order functions like

$$\frac{V_2}{V_1} = \frac{K(s^2 - 2B\omega_c s + 4C\omega_c^2)}{s^2 + 2B\omega_c s + 4C\omega_c^2} \qquad (7\text{-}37)$$

with one first-order function (if n is odd) like

$$\frac{V_2}{V_1} = \frac{K(s - 2C\omega_c)}{s + 2C\omega_c} \qquad (7\text{-}38)$$

where K is the stage gain and B and C are the Bessel filter coefficients of Appendix E.

The second-order transfer function (7-37) may be realized by the methods of Sections 7-2 and 7-3. In the first-order case, Eq. (7-38), we may use the circuits of Fig. 7-5 [19]. Figure 7-5(a) realizes Eq. (7-38) with $K = 1$ and

$$2C\omega_c = \frac{1}{RC_1} \qquad (7\text{-}39)$$

Thus we may choose the capacitance C_1 and obtain R by

$$R = \frac{1}{2\omega_c CC_1} \qquad (7\text{-}40)$$

Figure 7-5(b) realizes Eq. (7-38) with an inverting gain $K = -1$ and identical results Eqs. (7-39) and (7-40). In both cases a convenient value of the capacitance C_1 is approximately $10/f_c$ μF.

168 CONSTANT-TIME-DELAY FILTERS   CHAPTER 7

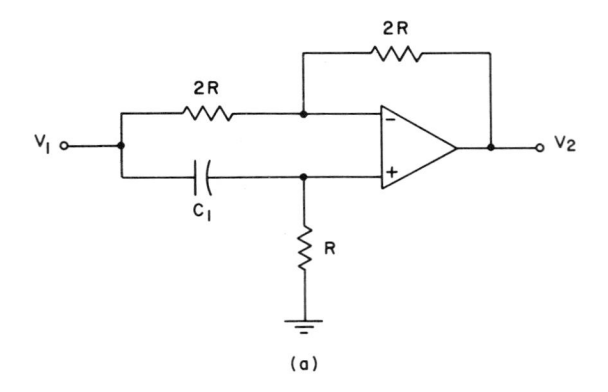

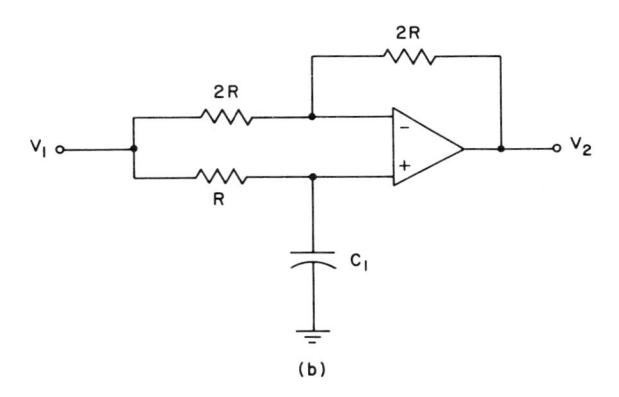

FIGURE 7-5 First-order all-pass filters with: (a) non-inverting and (b) inverting gain.

The design procedure for the all-pass, constant-time-delay filter is summarized in Section 7-9.

7-6 MFB ALL-PASS FILTER DESIGN SUMMARY

To design a second-order all-pass filter with a given phase shift ϕ_0 at a frequency f_0 Hz (or $\omega_0 = 2\pi f_0$ rad/s) and gain $K < 1$, perform the following steps.

SEC. 7-6 MFB ALL-PASS FILTER DESIGN SUMMARY

1. If $0 < \phi_0 < 180°$, calculate

$$a = \frac{(1-K)\{-1+\sqrt{1+[4K/(1-K)]\tan^2(\phi_0/2)}\}}{2K \tan(\phi_0/2)}$$

and if $-180° < \phi_0 < 0$, calculate

$$a = \frac{(1-K)\{-1-\sqrt{1+[4K/(1-K)]\tan^2(\phi_0/2)}\}}{2K \tan(\phi_0/2)}$$

2. Select a standard value of C_1 (preferably near $10/f_0$ μF) and calculate the resistance values given by

$$R_2 = \frac{2}{a\omega_0 C_1}$$

$$R_1 = \frac{1-K}{4K} R_2$$

$$R_3 = \frac{R_2}{K}$$

$$R_4 = \frac{R_2}{1-K}$$

For the special case $K = \frac{1}{2}$, use, for $0 < \phi_0 < 180°$,

$$a = \frac{-1+\sqrt{1+4\tan^2(\phi_0/2)}}{2\tan(\phi_0/2)}$$

and for $-180° < \phi_0 < 0$,

$$a = \frac{-1-\sqrt{1+4\tan^2(\phi_0/2)}}{2\tan(\phi_0/2)}$$

The resistances in this case are

$$R_2 = \frac{2}{a\omega_0 C_1}$$

$$R_1 = \frac{R_2}{4}$$

$$R_3 = R_4 = 2R_2$$

3. Select standard values of resistance as close as possible to the calculated values and construct the circuit in accordance with Fig. 7-6

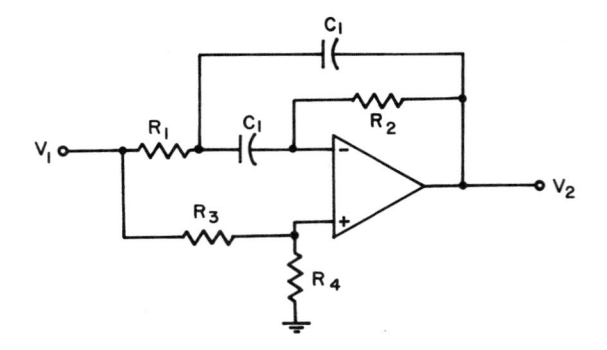

FIGURE 7-6 MFB all-pass filter circuit.

Comments

(a) For best performance, element values close to those selected and calculated should be used. The performance of the filter is unchanged if all the resistances are multiplied and the capacitances divided by a common factor

(b) The input impedance of the op amp should be at least $10R_{eq}$, where

$$R_{eq} = \frac{R_3 R_4}{R_3 + R_4}$$

The open-loop gain of the op amp should be at least 50 times the amplitude of the filter at f_0, and its slew rate (volts per microsecond) should be at least $\frac{1}{2}\omega_0 \times 10^{-6}$ times the peak-to-peak output voltage

(c) The gain is restricted to $K < 1$, and a reasonable value to choose is $K = \frac{1}{2}$. The gain and phase shift can be tuned by varying R_4 and R_2, respectively

The MFB all-pass filter was discussed in Section 7-2.

7-7 BIQUAD ALL-PASS FILTER DESIGN SUMMARY

To design a second-order all-pass filter with a given phase shift ϕ_0 at a frequency f_0 Hz (or $\omega_0 = 2\pi f_0$ rad/s) and a gain K, perform the following steps.

SEC. 7-7   BIQUAD ALL-PASS FILTER DESIGN   171

1. If $0 < \phi_0 < 180°$, calculate

$$a = \frac{-1 + \sqrt{1 + 4\tan^2(\phi_0/2)}}{2\tan(\phi_0/2)}$$

and if $-180° < \phi_0 < 0$, calculate

$$a = \frac{-1 - \sqrt{1 + 4\tan^2(\phi_0/2)}}{2\tan(\phi_0/2)}$$

2. Select a standard value of C_1 (preferably near $10/f_0$ μF) and calculate the resistance values given by

$$R_2 = \frac{1}{a\omega_0 C_1}$$

$$R_1 = \frac{R_2}{2K}$$

$$R_3 = \frac{R_2^2}{R_5}$$

$$R_4 = \frac{R_5}{K}$$

where R_5 is arbitrary and may be chosen to minimize the spread of the resistance values. A reasonable value is often $R_5 = R_2$, in which case we also have $R_3 = R_2$

3. Select standard values of resistance as close as possible to the calculated values and construct the circuit in accordance with Fig. 7-7

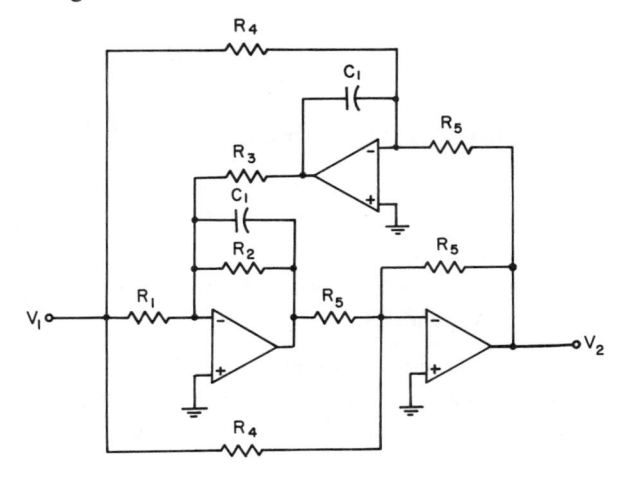

FIGURE 7-7 Biquad all-pass filter.

Comments

(a) Comments (a) and (b) for the MFB filter of Section 7-6 apply directly except that in (b), R_{eq} for each op amp is the resistance R_1 or R_5 connected to its inverting input terminal

(b) The gain is an inverting one of $-K$ ($K > 0$), given by $K = R_5/R_4$

(c) The circuit may be tuned by varying R_3 to adjust the phase shift and R_1 to adjust the gain. The parameter a is affected by changing R_2

The biquad all-pass filter was discussed in Section 7-3.

7-8 BESSEL (CONSTANT-TIME-DELAY) FILTER DESIGN SUMMARY

To design a Bessel, or constant-time-delay, filter with a given time delay τ seconds over a frequency range $0 < f < f_c$ Hz, or $0 < \omega < \omega_c = 2\pi f_c$ rad/s, gain K, and order $n = 2, 3, \ldots, 6$, perform the following steps.

1. If τ is given, find f_c from

$$f_c = \frac{0.15915}{\tau}$$

Alternatively, f_c may be specified, which determines

$$\tau = \frac{0.15915}{f_c}$$

2. Find the normalized coefficients B and C from the appropriate entry of Appendix E

3. Select the circuit, or circuits, to realize the filter, or its stages, from among the low-pass circuits summarized in Chapter 2. Construct the filter in accordance with Section 2-10, 2-11, 2-12, or 2-13

SEC. 7-9 ALL-PASS CONSTANT-TIME-DELAY FILTER   173

Comments

(a) See the comments for the low-pass circuit used in Chapter 2

(b) The time delay at f_c is slightly less than the given τ. If it is desirable for the time delay at f_c to be nearer to τ, it may be necessary to select a higher-order filter. For order $n = 2$ the time delay at f_c is 92.3% of τ, for $n = 3$ it is 99.6%, for $n = 4$ it is 99.99%, and for $n = 5$ or $n = 6$ it is virtually 100%

The Bessel filter was discussed in Section 7-4.

7-9 ALL-PASS CONSTANT-TIME-DELAY FILTER DESIGN SUMMARY

To design an all-pass constant-time-delay filter with a given time delay τ, gain K, and order $n = 2, 3, \ldots, 6$, perform the following steps.

1. Find f_c (Hz) or $\omega_c = 2\pi f_c$ (rad/s) from

$$f_c = \frac{0.15915}{\tau}$$

2. Find the normalized coefficients B and C from the appropriate entry of Appendix E

3. Select a standard value of capacitance C_1 (preferably near $10/f_c\ \mu F$)

4. In the case of a second-order filter, or a second-order stage of a higher-order filter, we may use (a) the MFB filter of Section 7-6 or (b) the biquad filter of Section 7-7. In case (a) (the MFB filter), the gain is noninverting and given by

$$K = \frac{C}{C + B^2}$$

and the resistance values are

$$R_2 = \frac{1}{B\omega_c C_1}$$

$$R_1 = \frac{B}{4\omega_c CC_1} = \frac{B^2}{4C}R_2$$

$$R_3 = \frac{R_2}{K}$$

$$R_4 = \frac{C}{B^2 K}R_2$$

In case (b) (the biquad filter) the gain is an inverting one of $-K$ ($K > 0$), and the resistance values are

$$R_2 = \frac{1}{2B\omega_c C_1}$$

$$R_1 = \frac{R_2}{2K}$$

$$R_3 = \frac{B^2 R_2^2}{R_5}$$

$$R_4 = \frac{R_5}{K}$$

where R_5 is arbitrary and may be chosen to minimize the resistance spread

5. If the filter is of odd order, then in addition to the second-order stages there will be one first-order stage corresponding to the first-order stage of Appendix E with coefficient C. A noninverting gain of 1 may be realized using the circuit of Fig. 7-5(a) and an inverting gain of -1 may be realized using the circuit of Fig. 7-5(b). In both cases

$$R = \frac{1}{2\omega_c CC_1}$$

6. Select standard values of resistance as nearly as possible to the calculated values and construct the filter, or its stages, in accordance with the appropriate one of Fig. 7-5, 7-6, or 7-7

Comments

See the comments of Section 7-6 or 7-7 as appropriate.

APPENDICES

The appendices that follow contain design data for normalized low-pass Butterworth, Chebyshev, inverse Chebyshev, elliptic, and Bessel filters. The transfer functions may be factored into second-order stage functions and, in the case of odd-order filters, a single first-order stage function. In the case of Butterworth, Chebyshev, and Bessel filters, a second-order stage function is of the form

$$\frac{V_2}{V_1} = \frac{KC\omega_c^2}{s^2 + B\omega_c s + C\omega_c^2}$$

In the case of inverse Chebyshev and elliptic filters, a second-order stage function is of the form

$$\frac{V_2}{V_1} = \frac{(KC/A)(s^2 + A\omega_c^2)}{s^2 + B\omega_c s + C\omega_c^2}$$

In the case of a first-order stage of Butterworth, Chebyshev, inverse Chebyshev, elliptic, or Bessel type, the stage function is

$$\frac{V_2}{V_1} = \frac{KC\omega_c}{s + C\omega_c}$$

In every case K is the stage gain and ω_c (rad/s) is the cutoff frequency of the filter, except for the Bessel filter, where the time delay at ω_c is approximately $1/\omega_c$ seconds.

The coefficients A, B, and C are those of the normalized case ($\omega_c = 1$), and are tabulated by rows in the appendices. Each row represents a stage and in the case of a first-order stage, only C is given. The data are arranged according to the order N of the filter.

The Butterworth and Chebyshev filter data are given in Appendix A, where the coefficients B and C are listed. In the case of the Chebyshev filter, passband ripple widths PRW of 0.1, 0.5, 1, 2, and 3 dB are available. The orders given are $N = 2, 3, \ldots, 10$.

The inverse Chebyshev data are given in Appendix B. The coefficients A, B, and C are listed for various values of minimum stopband loss MSL in dB. The characteristics WZ, WM, and KM are for use in tuning the filter stages, as discussed in Section 3-6. The orders listed are $N = 2, 3, \ldots, 10$.

The elliptic filter data are given in Appendix C, which is similar to Appendix B except for the addition of the normalized transition width TW. For each order N (2, 3, ..., 10), the data are listed according to the passband ripple widths PRW of 0.1, 0.5, 1, 2, or 3 dB.

Appendix D displays the normalized transition widths TW for the elliptic filter with a given passband ripple width PRW, minimum stopband loss MSL, and order N. For each value of PRW (0.1, 0.5, 1, 2, or 3 dB), TW is tabulated by rows according to MSL and by columns according to N.

The Bessel filter coefficients B and C are tabulated in Appendix E for $N = 2, 3, \ldots, 6$.

APPENDIX A

BUTTERWORTH AND CHEBYSHEV LOW-PASS FILTER DATA

N= 2

BUTTERWORTH

	B	C
	1.414214	1.000000

CHEBYSHEV

PRW	B	C
0.1	2.372356	3.314037
0.5	1.425625	1.516203
1.0	1.097734	1.102510
2.0	0.803816	0.823060
3.0	0.644900	0.707948

N= 3

BUTTERWORTH

B	C
1.000000	1.000000
–	1.000000

N= 3 (CONTINUED)

CHEBYSHEV

PRW	B	C
0.1	0.969406	1.689747
	–	0.969406
0.5	0.626456	1.142448
	–	0.626456
1.0	0.494171	0.994205
	–	0.494171
2.0	0.368911	0.886095
	–	0.368911
3.0	0.298620	0.839174
	–	0.298620

N= 4

BUTTERWORTH

B	C
0.765367	1.000000
1.847759	1.000000

177

APPENDIX A

N= 4 (CONTINUED)

CHEBYSHEV

PRW	B	C
0.1	0.528313 1.275460	1.330031 0.622925
0.5	0.350706 0.846680	1.063519 0.356412
1.0	0.279072 0.673739	0.986505 0.279398
2.0	0.209775 0.506440	0.928675 0.221568
3.0	0.170341 0.411239	0.903087 0.195980

N= 5

BUTTERWORTH

B	C
0.618034	1.000000
1.618034	1.000000
—	1.000000

CHEBYSHEV

PRW	B	C
0.1	0.333067 0.871982 —	1.194937 0.635920 0.538914
0.5	0.223926 0.586245 —	1.035784 0.476767 0.362320
1.0	0.178917 0.468410 —	0.988315 0.429298 0.289493
2.0	0.134922 0.353230 —	0.952167 0.393150 0.218308
3.0	0.109720 0.287250 —	0.936025 0.377009 0.177530

N= 6

BUTTERWORTH

B	C
0.517638	1.000000
1.414214	1.000000
1.931852	1.000000

N= 6 (CONTINUED)

CHEBYSHEV

PRW	B	C
0.1	0.229387 0.626696 0.856083	1.129387 0.696374 0.263361
0.5	0.155300 0.424288 0.579588	1.023023 0.590010 0.156997
1.0	0.124362 0.339763 0.464125	0.990732 0.557720 0.124707
2.0	0.093946 0.256666 0.350613	0.965952 0.532939 0.099926
3.0	0.076459 0.208890 0.285349	0.954830 0.521818 0.088805

N= 7

BUTTERWORTH

B	C
0.445042	1.000000
1.246980	1.000000
1.801938	1.000000
—	1.000000

CHEBYSHEV

PRW	B	C
0.1	0.167682 0.469834 0.678930 —	1.092446 0.753222 0.330217 0.376778
0.5	0.114006 0.319439 0.461602 —	1.016108 0.676884 0.253878 0.256170
1.0	0.091418 0.256147 0.370144 —	0.992679 0.653456 0.230450 0.205414
2.0	0.069133 0.193706 0.279913 —	0.974615 0.635391 0.212386 0.155340
3.0	0.056291 0.157725 0.227919 —	0.966483 0.627259 0.204254 0.126485

LOW-PASS FILTER DATA

N= 8

BUTTERWORTH

	C
0.390181	1.000000
1.111140	1.000000
1.662939	1.000000
1.961571	1.000000

CHEBYSHEV

PRW	B	C
0.1	0.127960	1.069492
	0.364400	0.798894
	0.545363	0.416210
	0.643300	0.145612
0.5	0.087240	1.011932
	0.248439	0.741334
	0.371815	0.353650
	0.438586	0.088052
1.0	0.070016	0.994141
	0.199390	0.723543
	0.298408	0.340859
	0.351997	0.070261
2.0	0.052985	0.980380
	0.150888	0.709782
	0.225820	0.327099
	0.266372	0.056501
3.0	0.043156	0.974173
	0.122899	0.703575
	0.183931	0.320892
	0.216961	0.050294

N= 9

BUTTERWORTH

B	C
0.347296	1.000000
1.000000	1.000000
1.532089	1.000000
1.879385	1.000000
–	1.000000

CHEBYSHEV

PRW	B	C
0.1	0.100876	1.054214
	0.290461	0.834368
	0.445012	0.497544
	0.545888	0.201345
	–	0.290461
0.5	0.068905	1.009211
	0.198405	0.789365
	0.303975	0.452541
	0.372880	0.156342
	–	0.198405

N= 9 (CONTINUED)

1.0	0.055335	0.995233
	0.159330	0.775386
	0.244108	0.438562
	0.299443	0.142364
	–	0.159330
2.0	0.041894	0.984398
	0.120630	0.764552
	0.184816	0.427727
	0.226710	0.131529
	–	0.120630
3.0	0.034130	0.979504
	0.098275	0.759658
	0.150565	0.422834
	0.184696	0.126636
	–	0.098275

N= 10

BUTTERWORTH

B	C
0.312869	1.000000
0.907981	1.000000
1.414214	1.000000
1.782013	1.000000
1.975377	1.000000

CHEBYSHEV

PRW	B	C
0.1	0.081577	1.043513
	0.236747	0.861878
	0.368742	0.567985
	0.464642	0.274093
	0.515059	0.092457
0.5	0.055799	1.007335
	0.161934	0.825700
	0.252219	0.531807
	0.317814	0.237915
	0.352300	0.056279
1.0	0.044829	0.996058
	0.130099	0.814423
	0.202633	0.520530
	0.255333	0.226637
	0.283039	0.045002
2.0	0.033952	0.987304
	0.098531	0.805669
	0.153466	0.511776
	0.193379	0.217883
	0.214362	0.036248
3.0	0.027664	0.983346
	0.080284	0.801711
	0.125045	0.507818
	0.157566	0.213926
	0.174663	0.032290

APPENDIX B

INVERSE CHEBYSHEV LOW-PASS FILTER DATA

N= 2

MSL	A	B	C	WZ	WM	KM
30.0	32.606961	1.413164	1.031123	5.710251	–	–
35.0	57.225240	1.413880	1.017625	7.564737	0.000000	1.000000
40.0	100.995000	1.414108	1.009950	10.049627	0.000000	1.000000
45.0	178.825129	1.414180	1.005608	13.372551	0.000000	1.000000
50.0	317.226185	1.414203	1.003157	17.810845	0.000000	1.000000
55.0	563.340436	1.414210	1.001777	23.734794	0.000000	1.000000
60.0	1000.999500	1.414213	1.000999	31.638576	0.000000	1.000000
65.0	1779.279129	1.414213	1.000562	42.181502	0.000000	1.000000
70.0	3163.277502	1.414213	1.000316	56.243022	0.000000	1.000000
75.0	5624.413163	1.414214	1.000178	74.996088	–	–

N= 3

MSL	A	B	C	WZ	WM	KM
30.0	5.976366	0.933370	1.058740 1.134320	2.444661	0.697349	1.122046
35.0	8.446367	0.955345	1.040981 1.089639	2.906263	0.700199	1.131688
40.0	12.075684	0.969938	1.028354 1.060226	3.475008	0.702266	1.138653
45.0	17.405719	0.979693	1.019514 1.040647	4.172016	0.703742	1.143590
50.0	25.231278	0.986247	1.013385 1.027516	5.023075	0.704782	1.147048

INVERSE CHEBYSHEV LOW-PASS FILTER DATA

N= 3 (CONTINUED)

MSL	A	B	C	WZ	WM	KM
55.0	36.719150	0.990669	1.009160 1.018665	6.059633	0.705507	1.149447
60.0	53.582108	0.993661	1.006260 1.012679	7.319980	0.706009	1.151103
65.0	78.334288	0.995690	1.004273 1.008621	8.850666	0.706356	1.152241
70.0	114.666032	0.997067	1.002915 1.005865	10.708223	0.706593	1.153021
75.0	167.994091	0.998004	1.001988 1.003992	12.961253	0.706756	1.153554
80.0	246.269218	0.998641	1.001355 1.002718	15.692967	0.706868	1.153919
85.0	361.161557	0.999074	1.000924 1.001851	19.004251	0.706944	1.154168
90.0	529.800560	0.999370	1.000630 1.001261	23.017397	0.706996	1.154337
95.0	777.328843	0.999571	1.000429 1.000859	27.880618	0.707031	1.154453
100.0	1140.650729	0.999708	1.000292 1.000585	33.773521	0.707055	1.154532

N= 4

MSL	A	B	C	WZ	WM	KM
30.0	2.951050 17.199978	0.630988 2.169970	1.061509 1.512100	1.717862 4.147286	0.824050 -	1.266126 -
35.0	3.719203 21.677103	0.664072 2.091600	1.048291 1.367632	1.928524 4.655868	0.827482 -	1.298081 -
40.0	4.748478 27.676159	0.689168 2.031494	1.037463 1.266740	2.179100 5.260814	0.830347 -	1.324115 -
45.0	6.124879 35.698408	0.708119 1.985895	1.028812 1.195116	2.474849 5.974815	0.832687 -	1.344887 -
50.0	7.963280 46.413396	0.722391 1.951490	1.022018 1.143607	2.821928 6.812738	0.834562 -	1.361205 -
55.0	10.417060 60.715075	0.733119 1.925605	1.016745 1.106189	3.227547 7.791988	0.836042 -	1.373875 -
60.0	13.690914 79.796493	0.741175 1.906160	1.012691 1.078796	3.700123 8.932888	0.837195 -	1.383628 -
65.0	18.057937 105.249368	0.747222 1.891564	1.009592 1.058624	4.249463 10.259111	0.838086 -	1.391086 -
70.0	23.882409 139.196880	0.751759 1.880612	1.007236 1.043702	4.886963 11.798173	0.838769 -	1.396762 -
75.0	31.650183 184.470785	0.755161 1.872397	1.005450 1.032626	5.625850 13.582002	0.839289 -	1.401065 -

APPENDIX B

N= 4 (CONTINUED)

MSL	A	B	C	WZ	WM	KM
80.0	42.009214 244.847640	0.757714 1.866236	1.004101 1.024385	6.481452 15.647608	0.839685 -	1.404319 -
85.0	55.823606 325.363818	0.759628 1.861615	1.003083 1.018240	7.471520 18.037844	0.839984 -	1.406774 -
90.0	74.245696 432.735628	0.761063 1.858150	1.002316 1.013653	8.616594 20.802299	0.840210 -	1.408624 -
95.0	98.812175 575.919560	0.762139 1.855551	1.001739 1.010224	9.940431 23.998324	0.840381 -	1.410015 -
100.0	131.572271 766.859392	0.762947 1.853602	1.001306 1.007659	11.470496 27.692226	0.840509 -	1.411062 -

N= 5

MSL	A	B	C	WZ	WM	KM
30.0	2.056891 5.385010	0.443052 1.697486	1.053952 1.542401 1.470208	1.434186 2.320563	0.881284 -	1.369787 -
35.0	2.422256 6.341549	0.476424 1.701078	1.045520 1.425895 1.356284	1.556360 2.518243	0.883984 -	1.424713 -
40.0	2.887037 7.558361	0.503909 1.696117	1.037939 1.334444 1.273011	1.699128 2.749247	0.886430 -	1.472746 -
45.0	3.476069 9.100466	0.526359 1.687442	1.031317 1.262886 1.210941	1.864422 3.016698	0.888601 -	1.513903 -
50.0	4.220752 11.050073	0.544578 1.677683	1.025653 1.206911 1.164000	2.054447 3.324165	0.890488 -	1.548614 -
55.0	5.160760 13.511044	0.559290 1.668189	1.020883 1.163078 1.128109	2.271731 3.675737	0.892102 -	1.577527 -
60.0	6.346154 16.614448	0.571127 1.659585	1.016917 1.128697 1.100438	2.519157 4.076082	0.893461 -	1.601373 -
65.0	7.840066 20.525559	0.580621 1.652099	1.013650 1.101681 1.078965	2.800012 4.530514	0.894594 -	1.620890 -
70.0	9.722051 25.452659	0.588220 1.645751	1.010978 1.080415 1.062219	3.118020 5.045063	0.895529 -	1.636764 -
75.0	12.092332 31.658137	0.594293 1.640460	1.008807 1.063649 1.049108	3.477403 5.626556	0.896294 -	1.649613 -

INVERSE CHEBYSHEV LOW-PASS FILTER DATA

N= 5 (CONTINUED)

MSL	A	B	C	WZ	WM	KM
80.0	15.077137 39.472456	0.599138 1.636104	1.007052 1.050413 1.038812	3.882929 6.282711	0.896917 –	1.659974 –
85.0	18.835415 49.311758	0.603001 1.632548	1.005637 1.039952 1.030707	4.339979 7.022233	0.897421 –	1.668302 –
90.0	23.567311 61.700021	0.606078 1.629663	1.004500 1.031676 1.024315	4.854617 7.854936	0.897828 –	1.674980 –
95.0	29.524814 77.296966	0.608528 1.627334	1.003589 1.025124 1.019266	5.433674 8.791869	0.898156 –	1.680324 –
100.0	37.025183 96.933187	0.610478 1.625461	1.002860 1.019933 1.015273	6.084832 9.845465	0.898418 –	1.684595 –

N= 6

MSL	A	B	C	WZ	WM	KM
30.0	1.670469 3.117138 23.266636	0.323814 1.246505 2.881119	1.044801 3.117138 2.490873	1.292466 1.765542 4.823550	0.913710 0.000000 –	1.440607 1.000000 –
35.0	1.884700 3.516898 26.250484	0.353422 1.297788 2.702736	1.039566 3.516898 2.130170	1.372844 1.875339 5.123523	0.915608 0.000000 –	1.514599 1.000000 –
40.0	2.148655 4.009446 29.926909	0.379124 1.333850 2.558246	1.034557 1.332266 1.870542	1.465829 2.002360 5.470549	0.917420 – –	1.582261 – –
45.0	2.472138 4.613072 34.432438	0.401242 1.358966 2.441615	1.029907 1.276766 1.679273	1.572303 2.147806 5.867916	0.919116 – –	1.642971 – –
50.0	2.867115 5.350110 39.933766	0.420134 1.376339 2.347446	1.025683 1.229877 1.535582	1.693256 2.313031 6.319317	0.920675 – –	1.696623 – –
55.0	3.348184 6.247796 46.634184	0.436173 1.388300 2.271273	1.021913 6.247796 1.425860	1.829804 2.499559 6.828923	0.922082 0.000000 –	1.743463 1.000000 –
60.0	3.933113 7.339289 54.781197	0.449720 1.396507 2.209511	1.018594 7.339289 1.340934	1.983208 2.709112 7.401432	0.923335 0.000000 –	1.783951 1.000000 –
65.0	4.643507 8.664902 64.675706	0.461114 1.402126 2.159312	1.015705 8.664902 1.274461	2.154880 2.943620 8.042121	0.924436 0.000000 –	1.818667 1.000000 –
70.0	5.505602 10.273593 76.683141	0.470666 1.405967 2.118418	1.013213 10.273593 1.221946	2.346402 3.205245 8.756891	0.925394 0.000000 –	1.848237 1.000000 –

APPENDIX B

N= 6 (CONTINUED)

MSL	A	B	C	WZ	WM	KM
75.0	6.551237	0.478650	1.011081	2.559538	0.926220	1.873284
	12.224775	1.408590	12.224775	3.496395	0.000000	1.000000
	91.246961	2.085037	1.180140	9.552328	–	–
80.0	7.819030	0.485310	1.009267	2.796253	0.926927	1.894406
	14.590508	1.410380	14.590508	3.819752	0.000000	1.000000
	108.905036	2.057740	1.146648	10.435758	–	–
85.0	9.355802	0.490853	1.007733	3.058726	0.927528	1.912151
	17.458164	1.411601	17.458164	4.178297	0.000000	1.000000
	130.309511	2.035384	1.119677	11.415319	–	–
90.0	11.218309	0.495461	1.006441	3.349374	0.928037	1.927013
	20.933650	1.412433	20.933650	4.575331	0.000000	1.000000
	156.250894	2.017051	1.097864	12.500036	–	–
95.0	13.475338	0.499286	1.005357	3.670877	0.928465	1.939430
	25.145323	1.413000	25.145323	5.014511	0.000000	1.000000
	187.687248	2.002000	1.080158	13.699900	–	–
100.0	16.210244	0.502458	1.004449	4.026195	0.928825	1.949781
	30.248727	1.413387	30.248727	5.499884	0.000000	1.000000
	225.779574	1.989633	1.065745	15.025963	–	–

N= 7

MSL	A	B	C	WZ	WM	KM
30.0	1.466461	0.245296	1.036751	1.210975	0.934324	1.489505
	2.280286	0.919245	2.280286	1.510062	0.457715	1.007335
	7.404042	2.293460	2.394066	2.721037	–	–
			1.880982			
35.0	1.608135	0.270458	1.033451	1.268123	0.935626	1.578291
	2.500582	0.983339	2.500582	1.581323	0.458909	1.008967
	8.119340	2.259519	2.132395	2.849446	–	–
			1.700559			
40.0	1.779084	0.293072	1.030156	1.333823	0.936911	1.661931
	2.766401	1.035073	2.766401	1.663250	0.460098	1.010588
	8.982447	2.215304	1.923194	2.997073	–	–
			1.564334			
45.0	1.983981	0.313234	1.026963	1.408538	0.938160	1.739353
	3.085007	1.076543	3.085007	1.756419	0.461260	1.012161
	10.016954	2.168355	1.755807	3.164957	–	–
			1.459103			
50.0	2.228395	0.331083	1.023936	1.492781	0.939352	1.810011
	3.465061	1.109648	3.465061	1.861467	0.462374	1.013656
	11.250982	2.122723	1.621400	3.354248	–	–
			1.376375			
55.0	2.518959	0.346786	1.021118	1.587123	0.940472	1.873758
	3.916876	1.136028	3.916876	1.979110	0.463424	1.015055
	12.718017	2.080416	1.512955	3.566233	–	–
			1.310435			
60.0	2.863553	0.360526	1.018529	1.692204	0.941509	1.930724
	4.452705	1.157050	4.452705	2.110143	0.464399	1.016344
	14.457842	2.042275	1.424993	3.802347	–	–
			1.257299			

INVERSE CHEBYSHEV LOW-PASS FILTER DATA

N= 7 (CONTINUED)

MSL	A	B	C	WZ	WM	KM
65.0	3.271520	0.372493	1.016181	1.808734	0.942457	1.981226
	5.087076	1.173828	5.087076	2.255455	0.465294	1.017517
	16.517632	2.008494	1.353273	4.064189	–	–
			1.214100			
70.0	3.753920	0.382874	1.014073	1.937504	0.943315	2.025698
	5.837188	1.187254	5.837188	2.416027	0.466105	1.018575
	18.953231	1.978927	1.294506	4.353531	–	–
			1.178729			
75.0	4.323833	0.391850	1.012195	2.079383	0.944084	2.064640
	6.723379	1.198033	6.723379	2.592948	0.466834	1.019519
	21.830674	1.953255	1.246133	4.672331	–	–
			1.149596			
80.0	4.996709	0.399588	1.010536	2.235332	0.944767	2.098578
	7.769674	1.206721	7.769674	2.787413	0.467483	1.020356
	25.227973	1.931092	1.206155	5.022746	–	–
			1.125485			
85.0	5.790793	0.406243	1.009078	2.406407	0.945370	2.128034
	9.004440	1.213753	9.004440	3.000740	0.468058	1.021093
	29.237233	1.912037	1.172995	5.407146	–	–
			1.105451			
90.0	6.727611	0.411954	1.007804	2.593764	0.945900	2.153513
	10.461153	1.219467	10.461153	3.234371	0.468563	1.021738
	33.967152	1.895702	1.145402	5.828134	–	–
			1.088749			
95.0	7.832563	0.416847	1.006696	2.798672	0.946363	2.175488
	12.179308	1.224132	12.179308	3.489887	0.469004	1.022300
	39.545965	1.881731	1.122380	6.288558	–	–
			1.074786			
100.0	9.135606	0.421033	1.005735	3.022517	0.946765	2.194393
	14.205486	1.227954	14.205486	3.769017	0.469389	1.022788
	46.124922	1.869802	1.103124	6.791533	–	–
			1.063087			

N= 8

MSL	A	B	C	WZ	WM	KM
30.0	1.344729	0.191520	1.030252	1.159625	0.948355	1.524116
	1.871070	0.695241	1.871070	1.367871	0.592152	1.019286
	4.190876	1.701624	2.147738	2.047163	–	–
	33.986884	3.644729	3.899919	5.829827	–	–
35.0	1.445893	0.212663	1.028112	1.202453	0.949252	1.624192
	2.011830	0.757039	2.011830	1.418390	0.593903	1.023825
	4.506154	1.752782	1.988226	2.122770	–	–
	36.543698	3.371832	3.242472	6.045138	–	–
40.0	1.566224	0.232122	1.025909	1.251489	0.950159	1.720434
	2.179260	0.810342	2.179260	1.476232	0.595701	1.028442
	4.881168	1.782018	1.847965	2.209337	–	–
	39.584967	3.146320	2.766030	6.291659	–	–
45.0	1.708248	0.249907	1.023708	1.306999	0.951063	1.811493
	2.376873	0.855972	2.376873	1.541711	0.597511	1.033035
	5.323788	1.796066	1.726276	2.307334	–	–
	43.174492	2.960075	2.411921	6.570730	–	–

APPENDIX B

N= 8 (CONTINUED)

MSL	A	B	C	WZ	WM	KM
50.0	1.874921	0.266056	1.021557	1.369277	0.951951	1.896525
	2.608784	0.894810	2.608784	1.615173	0.599300	1.037518
	5.843229	1.799951	1.621585	2.417277	-	-
	47.387017	2.805832	2.142957	6.883823	-	-
55.0	2.069706	0.280635	1.019489	1.438647	0.952809	1.975080
	2.879809	0.927732	2.879809	1.697000	0.601042	1.041823
	6.450280	1.797298	1.531968	2.539740	-	-
	52.310039	2.677576	1.934835	7.232568	-	-
60.0	2.296645	0.293729	1.017530	1.515469	0.953627	2.047002
	3.195575	0.955564	3.195575	1.787617	0.602712	1.045902
	7.157541	1.790653	1.455461	2.675358	-	-
	58.045733	2.570461	1.771221	7.618775	-	-
65.0	2.560447	0.305435	1.015695	1.600140	0.954398	2.112352
	3.562632	0.979058	3.562632	1.887494	0.604294	1.049720
	7.979685	1.781767	1.390224	2.824834	-	-
	64.713107	2.480621	1.640842	8.044446	-	-
70.0	2.866586	0.315857	1.013996	1.693099	0.955116	2.171348
	3.988596	0.998880	3.988596	1.997147	0.605776	1.053258
	8.933772	1.771815	1.334605	2.988942	-	-
	72.450491	2.404972	1.535738	8.511785	-	-
75.0	3.221411	0.325104	1.012435	1.794829	0.955779	2.224312
	4.482304	1.015606	4.482304	2.117145	0.607151	1.056506
	10.039593	1.761570	1.287166	3.168532	-	-
	81.418408	2.341046	1.450164	9.023215	-	-
80.0	3.632286	0.333281	1.011013	1.905856	0.956386	2.271634
	5.053999	1.029729	5.053999	2.248110	0.608416	1.059466
	11.320092	1.751522	1.246670	3.364534	-	-
	91.802909	2.286855	1.379898	9.581383	-	-
85.0	4.107733	0.340493	1.009726	2.026754	0.956938	2.313741
	5.715541	1.041670	5.715541	2.390720	0.609570	1.062145
	12.801834	1.741970	1.212065	3.577965	-	-
	103.819437	2.240788	1.321778	10.189182	-	-
90.0	4.657618	0.346838	1.008568	2.158151	0.957437	2.351072
	6.480656	1.051778	6.480656	2.545713	0.610617	1.064555
	14.515559	1.733079	1.182460	3.809929	-	-
	117.717289	2.201532	1.273402	10.849760	-	-
95.0	5.293347	0.352409	1.007531	2.300727	0.957885	2.384067
	7.365215	1.060349	7.365215	2.713893	0.611561	1.066713
	16.496821	1.724928	1.157104	4.061628	-	-
	133.784794	2.168007	1.232919	11.566538	-	-
100.0	6.028109	0.357291	1.006607	2.455221	0.958285	2.413151
	8.387572	1.067629	8.387572	2.896130	0.612408	1.068637
	18.786724	1.717541	1.135362	4.334366	-	-
	152.355294	2.139324	1.198882	12.343229	-	-

N= 9

MSL	A	B	C	WZ	WM	KM
30.0	1.265859	0.153342	1.025129	1.125104	0.958358	1.549282
	1.636918	0.540809	1.636918	1.279421	0.673629	1.030453
	2.971347	1.264005	1.915500	1.723759	-	-
	10.495060	2.882794	3.561353	3.239608	-	-
			2.321759			

INVERSE CHEBYSHEV LOW-PASS FILTER DATA

N= 9 (CONTINUED)

MSL	A	B	C	WZ	WM	KM
35.0	1.342006 1.735386 3.150086 11.126385	0.171132 0.595775 1.342932 2.813171	1.023699 1.735386 1.821003 3.109717 2.077498	1.158450 1.317341 1.774848 3.335624	0.958987 0.675411 – –	1.657999 1.037897 – –
40.0	1.431661 1.851322 3.360533 11.869701	0.187783 0.645110 1.403629 2.730744	1.022193 1.851322 1.731987 2.746892 1.890499	1.196520 1.360633 1.833176 3.445243	0.959632 0.677276 – –	1.764112 1.045594 – –
45.0	1.536314 1.986652 3.606185 12.737366	0.203274 0.689076 1.449444 2.645665	1.020653 1.986652 1.649737 2.454800 1.743802	1.239481 1.409486 1.898996 3.568945	0.960289 0.679190 – –	1.866121 1.053384 – –
50.0	1.657689 2.143605 3.891088 13.743667	0.217604 0.728021 1.483400 2.563406	1.019114 2.143605 1.574820 2.218494 1.626510	1.287513 1.464105 1.972584 3.707245	0.960946 0.681123 – –	1.962994 1.061131 – –
55.0	1.797777 2.324757 4.219916 14.905120	0.230790 0.762351 1.508087 2.486689	1.017598 2.324757 1.507306 2.026118 1.531296	1.340812 1.524715 2.054244 3.860715	0.961595 0.683046 – –	2.054073 1.068716 – –
60.0	1.958876 2.533078 4.598063 16.240766	0.242867 0.792495 1.525648 2.416663	1.016128 2.533078 1.446939 1.868444 1.453047	1.399599 1.591565 2.144309 4.029983	0.962228 0.684933 – –	2.138988 1.076047 – –
65.0	2.143625 2.771983 5.031723 17.772495	0.253879 0.818886 1.537807 2.353593	1.014717 2.771983 1.393271 1.738335 1.388090	1.464112 1.664927 2.243150 4.215744	0.962838 0.686762 – –	2.217592 1.083051 – –
70.0	2.355052 3.045385 5.528005 19.525406	0.263882 0.841941 1.545923 2.297277	1.013379 3.045385 1.345757 1.630273 1.333714	1.534618 1.745103 2.351171 4.418756	0.963420 0.688517 – –	2.289909 1.089676 – –
75.0	2.596622 3.357765 6.095041 21.528226	0.272934 0.862051 1.551055 2.247274	1.012119 3.357765 1.303818 1.539975 1.287875	1.611404 1.832421 2.468814 4.639852	0.963970 0.690185 – –	2.356088 1.095888 – –
80.0	2.872292 3.714243 6.742121 23.813771	0.281102 0.879575 1.554013 2.203041	1.010943 3.714243 1.266877 1.464099 1.249004	1.694784 1.927237 2.596559 4.879936	0.964485 0.691757 – –	2.416368 1.101669 – –
85.0	3.186581 4.120658 7.479850 26.419494	0.288449 0.894838 1.555412 2.164009	1.009853 4.120658 1.234386 1.400015 1.215877	1.785100 2.029940 2.734931 5.139990	0.964964 0.693226 – –	2.471052 1.107012 – –

APPENDIX B

N= 9 (CONTINUED)

MSL	A	B	C	WZ	WM	KM
90.0	3.544638	0.295042	1.008849	1.882721	0.965407	2.520482
	4.583672	0.908129	4.583672	2.140951	0.694592	1.111922
	8.320315	1.555714	1.205836	2.884496	-	-
	29.388092	2.129624	1.345642	5.421078	-	-
			1.187524			
95.0	3.952330	0.300945	1.007929	1.988047	0.965814	2.565021
	5.110870	0.919703	5.110870	2.260723	0.695852	1.116411
	9.277289	1.555265	1.180765	3.045864	-	-
	32.768210	2.099364	1.299316	5.724352	-	-
			1.163170			
100.0	4.416338	0.306219	1.007090	2.101508	0.966187	2.605041
	5.710892	0.929786	5.710892	2.389747	0.697009	1.120495
	10.366453	1.554321	1.158758	3.219698	-	-
	36.615235	2.072752	1.259701	6.051052	-	-
			1.142184			

N= 10

MSL	A	B	C	WZ	WM	KM
30.0	1.211647	0.125375	1.021100	1.100749	0.965741	1.568048
	1.488862	0.431569	1.488862	1.220189	0.730153	1.039853
	2.363992	0.961829	1.733024	1.537528	-	-
	5.734856	2.129638	3.045205	2.394756	-	-
	48.300453	4.437049	5.723554	6.949853	-	-
35.0	1.271202	0.140440	1.020117	1.127476	0.966190	1.683448
	1.562041	0.479168	1.562041	1.249817	0.731756	1.049860
	2.480186	1.042680	2.480186	1.574861	0.000000	1.000000
	6.016733	2.179198	2.779121	2.452903	-	-
	50.674492	4.074010	4.686988	7.118602	-	-
40.0	1.340795	0.154714	1.019061	1.157927	0.966657	1.797327
	1.647557	0.523011	1.647557	1.283572	0.733455	1.060328
	2.615967	1.110909	2.615967	1.617395	0.000000	1.000000
	6.346127	2.200062	2.544234	2.519152	-	-
	53.448731	3.771184	3.934239	7.310864	-	-
45.0	1.421369	0.168169	1.017964	1.192212	0.967137	1.908109
	1.746565	0.563139	1.746565	1.321577	0.735225	1.071061
	2.773170	1.167859	2.773170	1.665284	0.000000	1.000000
	6.727490	2.201288	2.339459	2.593740	-	-
	56.660668	3.518379	3.373200	7.527328	-	-
50.0	1.513998	0.180789	1.016848	1.230446	0.967627	2.014652
	1.860387	0.599658	1.860387	1.363960	0.737039	1.081879
	2.953895	1.214959	2.953895	1.718690	0.000000	1.000000
	7.165913	2.189641	2.162265	2.676922	-	-
	60.353186	3.306484	2.945529	7.768731	-	-
55.0	1.619914	0.192571	1.015729	1.272758	0.968118	2.116163
	1.990536	0.632732	1.990536	1.410864	0.738873	1.092625
	3.160544	1.253613	3.160544	1.777792	0.000000	1.000000
	7.667227	2.170004	2.009553	2.768976	-	-
	64.575381	3.127953	2.613125	8.035881	-	-
60.0	1.740524	0.203523	1.014623	1.319289	0.968605	2.212126
	2.138741	0.662560	2.138741	1.462443	0.740702	1.103167
	3.395861	1.285137	3.395861	1.842786	0.000000	1.000000
	8.238087	2.145805	1.878163	2.870207	-	-
	69.383315	2.976712	2.350394	8.329665	-	-

INVERSE CHEBYSHEV LOW-PASS FILTER DATA

N= 10 (CONTINUED)

MSL	A	B	C	WZ	WM	KM
65.0	1.877429	0.213663	1.013543	1.370193	0.969083	2.302238
	2.306968	0.689368	2.306968	1.518871	0.742507	1.113395
	3.662969	1.310713	3.662969	1.913889	0.000000	1.000000
	8.886071	2.119400	1.765130	2.980951	-	-
	74.840799	2.847915	2.139694	8.651058	-	-
70.0	2.032444	0.223017	1.012497	1.425638	0.969547	2.386370
	2.497450	0.713392	2.497450	1.580332	0.744271	1.123223
	3.965414	1.331377	3.965414	1.991335	0.000000	1.000000
	9.619776	2.092368	1.667802	3.101576	-	-
	81.020255	2.737693	1.968575	9.001125	-	-
75.0	2.207628	0.231617	1.011494	1.485809	0.969994	2.464521
	2.712714	0.734871	2.712714	1.647032	0.745977	1.132588
	4.307207	1.348015	1.302370	2.075381	-	-
	10.448939	2.065739	1.583869	3.232482	-	-
	88.003684	2.642945	1.828070	9.381028	-	-
80.0	2.405304	0.239499	1.010539	1.550904	0.970420	2.536795
	2.955617	0.754039	2.955617	1.719191	0.747615	1.141447
	4.692885	1.361376	1.270791	2.166307	-	-
	11.384563	2.040160	1.511348	3.374102	-	-
	95.883751	2.561171	1.711590	9.792025	-	-
85.0	2.628097	0.246704	1.009637	1.621141	0.970825	2.603371
	3.229382	0.771119	3.229382	1.797048	0.749175	1.149770
	5.127565	1.372081	5.127565	2.264413	0.000000	1.000000
	12.439063	2.016013	1.448555	3.526906	-	-
	104.765020	2.490340	1.614214	10.235479	-	-
90.0	2.878961	0.253271	1.008790	1.696750	0.971205	2.664485
	3.537642	0.786322	3.537642	1.880862	0.750653	1.157546
	5.617016	1.380644	1.216590	2.370024	-	-
	13.626430	1.993503	1.394067	3.691400	-	-
	114.765341	2.428792	1.532209	10.712859	-	-
95.0	3.161226	0.259245	1.007999	1.777984	0.971562	2.720408
	3.884487	0.799844	3.884487	1.970910	0.752042	1.164772
	6.167731	1.387483	1.193509	2.483492	-	-
	14.962422	1.972709	1.346682	3.868129	-	-
	126.017411	2.375157	1.462700	11.225748	-	-
100.0	3.478638	0.264666	1.007264	1.865111	0.971894	2.771441
	4.274520	0.811862	4.274520	2.067491	0.753342	1.171458
	6.787019	1.392939	6.787019	2.605191	0.000000	1.000000
	16.464766	1.953636	1.305388	4.057680	-	-
	138.670540	2.328301	1.403447	11.775846	-	-

APPENDIX C

ELLIPTIC LOW-PASS FILTER DATA

N= 2   PRW= 0.1 DB

MSL	TW	A	B	C	WZ	WM	KM
30.0	6.2301	104.047635	2.333416	3.328375	10.200374	0.708812	1.011579
35.0	8.6229	184.699634	2.350703	3.322508	13.590424	0.708066	1.011579

N= 2   PRW= 0.5 DB

MSL	TW	A	B	C	WZ	WM	KM
30.0	3.8087	45.741812	1.401021	1.532193	6.763269	0.711003	1.059254
35.0	5.3829	80.980111	1.411967	1.525381	8.998895	0.709300	1.059254
40.0	7.4892	143.631601	1.418001	1.521423	11.984640	0.708341	1.059254

N= 2   PRW= 1.0 DB

MSL	TW	A	B	C	WZ	WM	KM
30.0	3.0041	31.557423	1.077590	1.119700	5.617599	0.712776	1.122018
35.0	4.3034	55.747673	1.086571	1.112312	7.466436	0.710299	1.122018
40.0	6.0448	98.756423	1.091509	1.108065	9.937627	0.708904	1.122018
45.0	8.3738	175.233349	1.094250	1.105648	13.237573	0.708118	1.122018

N= 2   PRW= 2.0 DB

MSL	TW	A	B	C	WZ	WM	KM
30.0	2.2921	21.164003	0.787152	0.842554	4.600435	0.715610	1.258925
35.0	3.3454	37.258945	0.794608	0.834124	6.104011	0.711900	1.258925
40.0	4.7610	65.874745	0.798690	0.829314	8.116326	0.709806	1.258925
45.0	6.6570	116.758537	0.800950	0.826587	10.805486	0.708626	1.258925

ELLIPTIC LOW-PASS FILTER DATA

N= 2   PRW= 3.0 DB

MSL	TW	A	B	C	WZ	WM	KM
50.0	9.1918	207.242398	0.802210	0.825047	14.395916	0.707961	1.258925
30.0	1.9032	16.341050	0.629795	0.729928	4.042406	0.718179	1.412538
35.0	2.8202	28.679452	0.636577	0.720395	5.355320	0.713352	1.412538
40.0	4.0558	50.616359	0.640274	0.714975	7.114517	0.710625	1.412538
45.0	5.7129	89.623936	0.642316	0.711908	9.466992	0.709087	1.412538
50.0	7.9300	158.988974	0.643452	0.710178	12.609083	0.708221	1.412538

N= 3   PRW= 0.1 DB

MSL	TW	A	B	C	WZ	WM	KM
30.0	1.4550	7.860670	0.836642	1.653085 1.059440	2.803689	1.070199	1.371990
35.0	1.9331	11.298302	0.877898	1.665398 1.029675	3.361295	1.080161	1.393909
40.0	2.5195	16.345329	0.906579	1.673456 1.009982	4.042936	1.087438	1.409566
45.0	3.2359	23.754505	0.926380	1.678789 0.996828	4.873859	1.092640	1.420586
50.0	4.1088	34.630611	0.939991	1.682347 0.987987	5.884778	1.096304	1.428263
55.0	5.1703	50.595240	0.949318	1.684737 0.982018	7.113033	1.098858	1.433574
60.0	6.4597	74.028611	0.955699	1.686348 0.977977	8.603988	1.100625	1.437230
65.0	8.0246	108.424449	0.960057	1.687438 0.975235	10.412706	1.101842	1.439739
70.0	9.9228	158.910882	0.963032	1.688177 0.973373	12.605986	1.102677	1.441456

N= 3   PRW= 0.5 DB

MSL	TW	A	B	C	WZ	WM	KM
30.0	0.9232	4.750210	0.529742	1.148782 0.698719	2.179498	0.963106	1.662431
35.0	1.2753	6.725772	0.559750	1.147476 0.674892	2.593409	0.965612	1.699103
40.0	1.7115	9.629215	0.580638	1.146209 0.659093	3.103098	0.967604	1.725250
45.0	2.2479	13.893593	0.595071	1.145168 0.648525	3.727411	0.969105	1.743630

APPENDIX C

N= 3   PRW= 0.5 DB (CONTINUED)

MSL	TW	A	B	C	WZ	WM	KM
50.0	2.9043	20.154769	0.604996	1.144374 0.641415	4.489406	0.970199	1.756424
55.0	3.7049	29.346268	0.611800	1.143794 0.636612	5.417220	0.970978	1.765270
60.0	4.6793	42.838480	0.616455	1.143381 0.633359	6.545111	0.971525	1.771357
65.0	5.8635	62.642986	0.619635	1.143091 0.631152	7.914732	0.971906	1.775532
70.0	7.3012	91.712469	0.621805	1.142889 0.629652	9.576663	0.972169	1.778390
75.0	9.0454	134.380940	0.623286	1.142750 0.628632	11.592279	0.972349	1.780344

N= 3   PRW= 1.0 DB

MSL	TW	A	B	C	WZ	WM	KM
30.0	0.7325	3.816515	0.410567	1.016206 0.559558	1.953590	0.935416	1.914099
35.0	1.0366	5.351003	0.436466	1.009995 0.538016	2.313223	0.934568	1.964685
40.0	1.4162	7.608458	0.454520	1.005338 0.523721	2.758343	0.934166	2.000776
45.0	1.8851	10.925595	0.467003	1.001965 0.514155	3.305389	0.933982	2.026158
50.0	2.4606	15.797040	0.475592	0.999573 0.507718	3.974549	0.933903	2.043831
55.0	3.1640	22.949113	0.481481	0.997900 0.503369	4.790523	0.933870	2.056053
60.0	4.0212	33.448141	0.485510	0.996740 0.500423	5.783437	0.933859	2.064464
65.0	5.0639	48.859447	0.488263	0.995940 0.498423	6.989953	0.933856	2.070234
70.0	6.3305	71.480724	0.490143	0.995391 0.497065	8.454627	0.933857	2.074184
75.0	7.8679	104.684611	0.491425	0.995014 0.496141	10.231550	0.933858	2.076884
80.0	9.7327	153.421520	0.492299	0.994757 0.495512	12.386344	0.933860	2.078727

N= 3   PRW= 2.0 DB

MSL	TW	A	B	C	WZ	WM	KM
30.0	0.5568	3.040224	0.297467	0.924369 0.427424	1.743624	0.918241	2.344205

ELLIPTIC LOW-PASS FILTER DATA

N= 3   PRW= 2.0 DB (CONTINUED)

MSL	TW	A	B	C	WZ	WM	KM
35.0	0.8144	4.205740	0.319529	0.913153 0.408146	2.050790	0.913783	2.421291
40.0	1.1392	5.923501	0.334950	0.904987 0.395352	2.433824	0.910771	2.476448
45.0	1.5433	8.449743	0.345631	0.899178 0.386791	2.906844	0.908737	2.515315
50.0	2.0414	12.161171	0.352985	0.895107 0.381030	3.487287	0.907362	2.542414
55.0	2.6519	17.611155	0.358031	0.892281 0.377139	4.196565	0.906431	2.561170
60.0	3.3973	25.612252	0.361485	0.890331 0.374503	5.060855	0.905800	2.574087
65.0	4.3051	37.357363	0.363845	0.888991 0.372715	6.112067	0.905371	2.582953
70.0	5.4089	54.597585	0.365457	0.888072 0.371499	7.389018	0.905080	2.589023
75.0	6.7494	79.903287	0.366556	0.887444 0.370673	8.938864	0.904882	2.593173
80.0	8.3761	117.047331	0.367306	0.887015 0.370111	10.818841	0.904747	2.596007

N= 3   PRW= 3.0 DB

MSL	TW	A	B	C	WZ	WM	KM
30.0	0.4581	2.638395	0.234092	0.888381 0.352928	1.624314	0.913244	2.754264
35.0	0.6878	3.611302	0.253951	0.873882 0.335021	1.900343	0.906457	2.859860
40.0	0.9802	5.047821	0.267872	0.863370 0.323142	2.246736	0.901730	2.935694
45.0	1.3460	7.162265	0.277531	0.855913 0.315198	2.676241	0.898468	2.989264
50.0	1.7986	10.269942	0.284188	0.850696 0.309854	3.204675	0.896228	3.026677
55.0	2.3547	14.834210	0.288759	0.847078 0.306246	3.851520	0.894694	3.052602
60.0	3.0347	21.535572	0.291888	0.844584 0.303803	4.640644	0.893645	3.070471
65.0	3.8638	31.373152	0.294028	0.842872 0.302145	5.601174	0.892930	3.082741
70.0	4.8725	45.813648	0.295488	0.841699 0.301019	6.768578	0.892441	3.091145
75.0	6.0981	67.010016	0.296485	0.840897 0.300253	8.185965	0.892109	3.096892

APPENDIX C

N= 3   PRW= 3.0 DB (CONTINUED)

MSL	TW	A	B	C	WZ	WM	KM
80.0	7.5859	98.122451	0.297165	0.840349	9.905678	0.891882	3.100818
		–	–	0.299732	–	–	–
85.0	9.3907	143.789547	0.297629	0.839975	11.991228	0.891727	3.103496
		–	–	0.299377	–	–	–

N= 4   PRW= 0.1 DB

MSL	TW	A	B	C	WZ	WM	KM
30.0	0.5304	2.644233	1.392323	0.854559	1.626110	–	–
		12.746572	0.355395	1.261686	3.570234	1.088613	2.901587
35.0	0.7122	3.338443	1.367043	0.791865	1.827141	–	–
		16.855287	0.393949	1.278289	4.105519	1.090139	2.707970
40.0	0.9298	4.269241	1.346123	0.746910	2.066214	–	–
		22.327394	0.424844	1.290982	4.725187	1.090914	2.576668
45.0	1.1875	5.514373	1.329473	0.714377	2.348270	–	–
		29.619871	0.449171	1.300624	5.442414	1.091271	2.485214
50.0	1.4907	7.177749	1.316503	0.690648	2.679132	–	–
		39.341242	0.468081	1.307916	6.272260	1.091406	2.420276
55.0	1.8457	9.398140	1.306525	0.673229	3.065639	–	–
		52.302514	0.482645	1.313415	7.232048	1.091431	2.373514
60.0	2.2597	12.360770	1.298912	0.660377	3.515789	–	–
		69.584898	0.493783	1.317554	8.341756	1.091404	2.339490
65.0	2.7415	16.312770	1.293131	0.650858	4.038907	–	–
		92.630031	0.502259	1.320666	9.624450	1.091359	2.314543
70.0	3.3009	21.583802	1.288758	0.643787	4.645837	–	–
		123.360245	0.508685	1.323004	11.106766	1.091310	2.296148
75.0	3.9495	28.613552	1.285458	0.638521	5.349164	–	–
		164.338921	0.513543	1.324759	12.819474	1.091266	2.282526
80.0	4.7008	37.988418	1.282972	0.634594	6.163475	–	–
		218.984367	0.517208	1.326076	14.798120	1.091228	2.272407
85.0	5.5704	50.490396	1.281101	0.631660	7.105659	–	–
		291.854770	0.519970	1.327065	17.083757	1.091196	2.264871
90.0	6.5763	67.162395	1.279695	0.629467	8.195267	–	–
		389.028956	0.522047	1.327806	19.723817	1.091171	2.259251
95.0	7.7394	89.395002	1.278638	0.627826	9.454893	–	–
		518.612077	0.523609	1.328363	22.773056	1.091152	2.255052
100.0	9.0838	119.043513	1.277845	0.626597	10.910706	–	–
		691.417752	0.524783	1.328780	26.294824	1.091136	2.251913

N= 4   PRW= 0.5 DB

MSL	TW	A	B	C	WZ	WM	KM
30.0	0.3244	1.948438	0.946079	0.528346	1.395865	–	–
		8.563850	0.219820	1.057881	2.926406	1.013415	4.139729

ELLIPTIC LOW-PASS FILTER DATA

N= 4   PRW= 0.5 DB (CONTINUED)

MSL	TW	A	B	C	WZ	WM	KM
35.0	0.4611	2.399620 11.287390	0.924791 0.248639	0.481297 1.060041	1.549071 3.359671	– 1.011353	– 3.792233
40.0	0.6284	3.009057 14.910257	0.907057 0.271917	0.447762 1.061374	1.734663 3.861380	– 1.009382	– 3.560300
45.0	0.8298	3.827702 19.734952	0.892887 0.290341	0.423616 1.062190	1.956451 4.442404	– 1.007647	– 3.400558
50.0	1.0692	4.923892 26.164053	0.881820 0.304712	0.406076 1.062688	2.218985 5.115081	– 1.006194	– 3.288041
55.0	1.3518	6.389088 34.733900	0.873293 0.315804	0.393244 1.062993	2.527665 5.893547	– 1.005016	– 3.207488
60.0	1.6832	8.345520 46.159374	0.866779 0.324301	0.383801 1.063181	2.888861 6.794069	– 1.004081	– 3.149127
65.0	2.0704	10.956388 61.393548	0.861829 0.330774	0.376821 1.063297	3.310044 7.835403	– 1.003352	– 3.106471
70.0	2.5213	14.439481 81.707192	0.858083 0.335686	0.371644 1.063371	3.799932 9.039203	– 1.002788	– 3.075090
75.0	3.0454	19.085345 108.794786	0.855254 0.339401	0.367793 1.063419	4.368678 10.430474	– 1.002355	– 3.051892
80.0	3.6533	25.281516 144.915857	0.853123 0.342205	0.364924 1.063450	5.028073 12.038100	– 1.002026	– 3.034680
85.0	4.3578	33.544853 193.083468	0.851519 0.344318	0.362782 1.063470	5.791792 13.895448	– 1.001776	– 3.021877
90.0	5.1735	44.564654 257.315590	0.850312 0.345908	0.361181 1.063484	6.675676 16.041059	– 1.001586	– 3.012332
95.0	6.1173	59.260142 342.970178	0.849406 0.347104	0.359984 1.063494	7.698061 18.519454	– 1.001444	– 3.005207
100.0	7.2087	78.857195 457.192434	0.848725 0.348003	0.359089 1.063501	8.880157 21.382059	– 1.001336	– 2.999882

N= 4   PRW= 1.0 DB

MSL	TW	A	B	C	WZ	WM	KM
30.0	0.2504	1.720855 7.160442	0.763449 0.165852	0.435190 1.004627	1.311814 2.675900	0.191780 0.993190	1.003361 5.226913
35.0	0.3686	2.089586 9.422090	0.744516 0.190508	0.392257 1.001468	1.445540 3.069542	0.209495 0.989471	1.006097 4.727404
40.0	0.5155	2.590652 12.427651	0.728581 0.210563	0.361768 0.998545	1.609550 3.525287	0.218185 0.986181	1.008602 4.398088
45.0	0.6941	3.265999 16.428030	0.715777 0.226506	0.339888 0.996016	1.807207 4.053151	0.222754 0.983412	1.010713 4.173233
50.0	0.9082	4.172024 21.756998	0.705743 0.238979	0.324038 0.993917	2.042553 4.664440	0.225268 0.981158	1.012418 4.015834

APPENDIX C

N= 4   PRW= 1.0 DB (CONTINUED)

MSL	TW	A	B	C	WZ	WM	KM
55.0	1.1620	5.384334	0.697995	0.312468	2.320417	0.226700	1.013761
		28.859108	0.248626	0.992225	5.372067	0.979364	3.903655
60.0	1.4608	7.004065	0.692067	0.303969	2.646519	0.227541	1.014802
		38.326802	0.256026	0.990886	6.190864	0.977958	3.822648
65.0	1.8107	9.166332	0.687559	0.297695	3.027595	0.228049	1.015601
		50.949839	0.261669	0.989843	7.137916	0.976870	3.763583
70.0	2.2190	12.051506	0.684144	0.293047	3.471528	0.228364	1.016211
		67.781179	0.265954	0.989038	8.232933	0.976035	3.720209
75.0	2.6941	15.900256	0.681565	0.289593	3.987512	0.228564	1.016674
		90.224822	0.269196	0.988422	9.498675	0.975398	3.688187
80.0	3.2459	21.033627	0.679620	0.287020	4.586243	0.228695	1.017024
		120.152917	0.271645	0.987952	10.961429	0.974913	3.664453
85.0	3.8858	27.879824	0.678157	0.285101	5.280135	0.228782	1.017289
		160.061936	0.273490	0.987595	12.651559	0.974547	3.646810
90.0	4.6271	37.009926	0.677056	0.283668	6.083578	0.228841	1.017488
		213.280914	0.274880	0.987326	14.604140	0.974270	3.633666
95.0	5.4851	49.185530	0.676229	0.282596	7.013240	0.228882	1.017638
		284.249169	0.275924	0.987122	16.859691	0.974061	3.623858
100.0	6.4776	65.422353	0.675607	0.281794	8.088409	0.228911	1.017751
		378.887033	0.276709	0.986969	19.465021	0.973904	3.616530

N= 4   PRW= 2.0 DB

MSL	TW	A	B	C	WZ	WM	KM
30.0	0.1828	1.522811	0.584170	0.369149	1.234022	0.354091	1.059595
		5.905875	0.114884	0.969903	2.430201	0.980165	7.188332
35.0	0.2821	1.817037	0.568240	0.328084	1.347975	0.343861	1.069770
		7.757221	0.135161	0.961802	2.785179	0.974729	6.381192
40.0	0.4084	2.220728	0.554550	0.299065	1.490211	0.335231	1.077701
		10.213846	0.151848	0.954816	3.195911	0.970014	5.859200
45.0	0.5643	2.767785	0.543427	0.278333	1.663666	0.328291	1.083789
		13.480726	0.165216	0.949032	3.671611	0.966092	5.507560
50.0	0.7528	3.503937	0.534654	0.263374	1.871881	0.322851	1.088423
		17.830389	0.175729	0.944375	4.222604	0.962924	5.263771
55.0	0.9780	4.490633	0.527852	0.252489	2.119111	0.318650	1.091930
		23.625672	0.183888	0.940696	4.860625	0.960417	5.091229
60.0	1.2444	5.810197	0.522634	0.244514	2.410435	0.315435	1.094577
		31.349996	0.190162	0.937831	5.599107	0.958460	4.967266
65.0	1.5574	7.572710	0.518659	0.238639	2.751856	0.312990	1.096570
		41.647692	0.194955	0.935621	6.453502	0.956950	4.877218
70.0	1.9236	9.925194	0.515644	0.234294	3.150428	0.311138	1.098069
		55.377748	0.198598	0.933929	7.441623	0.955793	4.811274
75.0	2.3505	13.063886	0.513365	0.231068	3.614400	0.309739	1.099195
		73.685467	0.201358	0.932641	8.584024	0.954911	4.762690

ELLIPTIC LOW-PASS FILTER DATA

N= 4   PRW= 2.0 DB (CONTINUED)

MSL	TW	A	B	C	WZ	WM	KM
80.0	2.8470	17.250603 98.097999	0.511646 0.203443	0.228668 0.931663	4.153384 9.904443	0.308685 0.954241	1.100041 4.726736
85.0	3.4232	22.834581 130.651730	0.510351 0.205015	0.226879 0.930924	4.778554 11.430299	0.307891 0.953735	1.100676 4.700039
90.0	4.0913	30.281612 174.062147	0.509377 0.206199	0.225543 0.930366	5.502873 13.193261	0.307294 0.953353	1.101152 4.680166
95.0	4.8650	40.212901 231.950398	0.508644 0.207090	0.224545 0.929946	6.341364 15.229918	0.306845 0.953065	1.101510 4.665346
100.0	5.7603	53.456882 309.145311	0.508094 0.207759	0.223798 0.929630	7.311421 17.582529	0.306508 0.952848	1.101778 4.654279

N= 4   PRW= 3.0 DB

MSL	TW	A	B	C	WZ	WM	KM
30.0	0.1454	1.416828 5.211857	0.480516 0.086865	0.344290 0.958043	1.190306 2.282949	0.416108 0.976000	1.148279 9.215829
35.0	0.2331	1.669333 6.837923	0.466754 0.104384	0.302707 0.947207	1.292027 2.614942	0.396390 0.969543	1.164761 8.052399
40.0	0.3466	2.018835 8.992794	0.454659 0.118962	0.273432 0.937876	1.420857 2.998799	0.381157 0.963929	1.177339 7.312711
45.0	0.4884	2.494794 11.856156	0.444717 0.130726	0.252593 0.930159	1.579492 3.443277	0.369519 0.959254	1.186857 6.820298
50.0	0.6615	3.137050 15.666825	0.436825 0.140021	0.237603 0.923952	1.771172 3.958134	0.360689 0.955473	1.194025 6.481779
55.0	0.8692	3.999233 20.742655	0.430681 0.147260	0.226725 0.919055	1.999808 4.554411	0.354016 0.952478	1.199412 6.243645
60.0	1.1160	5.153290 27.507045	0.425956 0.152839	0.218772 0.915244	2.270086 5.244716	0.348987 0.950140	1.203454 6.073314
65.0	1.4067	6.695495 36.524247	0.422350 0.157107	0.212923 0.912307	2.587565 6.043529	0.345202 0.948334	1.206486 5.949988
70.0	1.7475	8.754499 48.546435	0.419612 0.160356	0.208602 0.910059	2.958800 6.967527	0.342358 0.946950	1.208759 5.859890
75.0	2.1453	11.502055 64.576448	0.417540 0.162819	0.205398 0.908348	3.391468 8.035947	0.340221 0.945895	1.210463 5.793628
80.0	2.6085	15.167355 85.951439	0.415977 0.164681	0.203016 0.907051	3.894529 9.271000	0.338617 0.945095	1.211741 5.744656
85.0	3.1465	20.056141 114.454417	0.414799 0.166086	0.201242 0.906070	4.478408 10.698337	0.337413 0.944489	1.212698 5.708330
90.0	3.7706	26.576214 152.462975	0.413912 0.167144	0.199917 0.905330	5.155212 12.347590	0.336510 0.944032	1.213417 5.681309
95.0	4.4936	35.271452 203.147619	0.413246 0.167940	0.198928 0.904772	5.938977 14.252986	0.335832 0.943687	1.213955 5.661170
100.0	5.3307	46.867181 270.736298	0.412745 0.168539	0.198188 0.904352	6.845961 16.454066	0.335323 0.943427	1.214359 5.646136

APPENDIX C

N= 5   PRW= 0.1 DB

MSL	TW	A	B	C	WZ	WM	KM
30.0	0.2258	1.591551 3.167724 −	0.771928 0.160923 −	0.890843 1.116892 0.761237	1.261567 1.779810 −	0.387613 1.044035 −	1.010002 4.305536 −
35.0	0.3129	1.839358 3.854820 −	0.802917 0.188718 −	0.840535 1.130878 0.709287	1.356229 1.963370 −	0.432079 1.048092 −	1.020322 4.034458 −
40.0	0.4176	2.158236 4.720464 −	0.823821 0.213351 −	0.799278 1.142739 0.670590	1.469094 2.172663 −	0.456433 1.051512 −	1.030967 3.848484 −
45.0	0.5414	2.565298 5.810828 −	0.837993 0.234653 −	0.765949 1.152644 0.641324	1.601655 2.410566 −	0.471212 1.054386 −	1.040945 3.716964 −
50.0	0.6858	3.082284 7.184003 −	0.847685 0.252732 −	0.739251 1.160819 0.618939	1.755644 2.680299 −	0.480793 1.056790 −	1.049830 3.621750 −
55.0	0.8527	3.736758 8.913123 −	0.854387 0.267857 −	0.717966 1.167507 0.601672	1.933070 2.985485 −	0.487300 1.058789 −	1.057493 3.551539 −
60.0	1.0444	4.563587 11.090273 −	0.859080 0.280371 −	0.701041 1.172942 0.588267	2.136255 3.330206 −	0.491878 1.060440 −	1.063966 3.498999 −
65.0	1.2632	5.606809 13.831394 −	0.862410 0.290633 −	0.687602 1.177336 0.577809	2.367870 3.719058 −	0.495186 1.061796 −	1.069351 3.459217 −
70.0	1.5122	6.921985 17.282461 −	0.864805 0.298993 −	0.676936 1.180875 0.569618	2.630967 4.157218 −	0.497625 1.062903 −	1.073785 3.428808 −
75.0	1.7946	8.579153 21.627255 −	0.866549 0.305767 −	0.668473 1.183716 0.563184	2.929019 4.650511 −	0.499453 1.063802 −	1.077405 3.405385 −
80.0	2.1140	10.666566 27.097157 −	0.867835 0.311233 −	0.661760 1.185992 0.558117	3.265971 5.205493 −	0.500841 1.064530 −	1.080344 3.387233 −
85.0	2.4749	13.295383 33.983449 −	0.868795 0.315630 −	0.656433 1.187811 0.554121	3.646283 5.829533 −	0.501904 1.065117 −	1.082718 3.373095 −
90.0	2.8818	16.605608 42.652878 −	0.869517 0.319156 −	0.652205 1.189264 0.550964	4.074998 6.530917 −	0.502725 1.065588 −	1.084628 3.362041 −
95.0	3.3403	20.773501 53.567063 −	0.870066 0.321979 −	0.648851 1.190423 0.548468	4.557796 7.318952 −	0.503362 1.065967 −	1.086162 3.353369 −
100.0	3.8564	26.021148 67.307566 −	0.870487 0.324236 −	0.646187 1.191346 0.546491	5.101093 8.204119 −	0.503860 1.066270 −	1.087390 3.346550 −

ELLIPTIC LOW-PASS FILTER DATA

N= 5   PRW= 0.5 DB

MSL	TW	A	B	C	WZ	WM	KM
30.0	0.1291	1.333847	0.510799	0.732427	1.154923	0.603417	1.108732
		2.421823	0.093582	1.024337	1.556221	1.006763	6.281208
		−	−	0.543429	−	−	−
35.0	0.1929	1.501984	0.535157	0.681055	1.225555	0.597344	1.140420
		2.913728	0.113861	1.026958	1.706965	1.006710	5.802979
		−	−	0.501433	−	−	−
40.0	0.2726	1.722931	0.551409	0.639213	1.312605	0.590434	1.168369
		3.534235	0.132172	1.028998	1.879956	1.006550	5.478608
		−	−	0.470007	−	−	−
45.0	0.3695	2.008726	0.562257	0.605620	1.417295	0.583885	1.192361
		4.316373	0.148211	1.030568	2.077588	1.006357	5.250957
		−	−	0.446162	−	−	−
50.0	0.4847	2.374727	0.569532	0.578860	1.541015	0.578116	1.212577
		5.301775	0.161946	1.031769	2.302558	1.006170	5.087001
		−	−	0.427884	−	−	−
55.0	0.6198	2.840495	0.574452	0.557630	1.685377	0.573219	1.229384
		6.542906	0.173509	1.032685	2.557910	1.006005	4.966526
		−	−	0.413763	−	−	−
60.0	0.7766	3.430868	0.577813	0.540817	1.852260	0.569152	1.243218
		8.105853	0.183118	1.033385	2.847078	1.005868	4.876594
		−	−	0.402789	−	−	−
65.0	0.9572	4.177303	0.580138	0.527513	2.043845	0.565818	1.254517
		10.073844	0.191026	1.033920	3.173932	1.005757	4.808617
		−	−	0.394222	−	−	−
70.0	1.1638	5.119559	0.581766	0.516985	2.262644	0.563109	1.263693
		12.551680	0.197483	1.034331	3.542835	1.005668	4.756718
		−	−	0.387508	−	−	−
75.0	1.3992	6.307818	0.582922	0.508652	2.511537	0.560922	1.271109
		15.671314	0.202725	1.034648	3.958701	1.005598	4.716778
		−	−	0.382232	−	−	−
80.0	1.6664	7.805364	0.583753	0.502053	2.793808	0.559162	1.277081
		19.598876	0.206960	1.034894	4.427062	1.005543	4.685844
		−	−	0.378077	−	−	−
85.0	1.9692	9.691945	0.584359	0.496826	3.113189	0.557751	1.281877
		24.543523	0.210370	1.035085	4.954142	1.005500	4.661763
		−	−	0.374799	−	−	−
90.0	2.3113	12.068032	0.584806	0.492683	3.473907	0.556623	1.285720
		30.768579	0.213108	1.035233	5.546943	1.005467	4.642939
		−	−	0.372209	−	−	−
95.0	2.6974	15.060159	0.585139	0.489398	3.880742	0.555721	1.288793
		38.605550	0.215301	1.035350	6.213336	1.005440	4.628177
		−	−	0.370160	−	−	−
100.0	3.1326	18.827677	0.585390	0.486793	4.339087	0.555002	1.291248
		48.471805	0.217055	1.035441	6.962170	1.005420	4.616570
		−	−	0.368538	−	−	−

APPENDIX C

N= 5   PRW= 1.0 DB

MSL	TW	A	B	C	WZ	WM	KM
30.0	0.0955	1.248003	0.401840	0.694011	1.117141	0.668571	1.220515
		2.157451	0.067616	1.001756	1.468826	0.997757	7.963080
		-	-	0.450563	-	-	-
35.0	0.1495	1.386814	0.423679	0.640619	1.177631	0.650340	1.267985
		2.579616	0.084434	1.000399	1.606118	0.996161	7.285193
		-	-	0.413282	-	-	-
40.0	0.2187	1.572033	0.438213	0.597139	1.253808	0.634552	1.309012
		3.112714	0.099841	0.998891	1.764288	0.994595	6.830033
		-	-	0.385344	-	-	-
45.0	0.3042	1.813923	0.447859	0.562269	1.346820	0.621289	1.343778
		3.785055	0.113472	0.997381	1.945522	0.993158	6.512853
		-	-	0.364129	-	-	-
50.0	0.4072	2.125573	0.454277	0.534533	1.457934	0.610347	1.372818
		4.632398	0.125225	0.995962	2.152301	0.991893	6.285581
		-	-	0.347860	-	-	-
55.0	0.5292	2.523683	0.458574	0.512562	1.588610	0.601423	1.396815
		5.699838	0.135169	0.994683	2.387433	0.990813	6.119206
		-	-	0.335288	-	-	-
60.0	0.6716	3.029508	0.461475	0.495191	1.740548	0.594198	1.416482
		7.044205	0.143462	0.993566	2.654092	0.989909	5.995358
		-	-	0.325518	-	-	-
65.0	0.8364	3.670012	0.463456	0.481464	1.915728	0.588380	1.432495
		8.737034	0.150305	0.992611	2.955856	0.989161	5.901940
		-	-	0.317890	-	-	-
70.0	1.0257	4.479318	0.464825	0.470615	2.116440	0.583712	1.445467
		10.868622	0.155903	0.991808	3.296759	0.988551	5.830733
		-	-	0.311912	-	-	-
75.0	1.2419	5.500530	0.465783	0.462037	2.345321	0.579975	1.455934
		13.552329	0.160454	0.991142	3.681349	0.988055	5.776001
		-	-	0.307216	-	-	-
80.0	1.4880	6.788034	0.466462	0.455251	2.605386	0.576990	1.464351
		16.931122	0.164136	0.990593	4.114744	0.987655	5.733650
		-	-	0.303517	-	-	-
85.0	1.7671	8.410398	0.466950	0.449878	2.900069	0.574608	1.471104
		21.184932	0.167103	0.990146	4.602709	0.987333	5.700706
		-	-	0.300599	-	-	-
90.0	2.0830	10.454019	0.467306	0.445623	3.233267	0.572709	1.476510
		26.540291	0.169486	0.989783	5.151727	0.987074	5.674969
		-	-	0.298294	-	-	-
95.0	2.4399	13.027727	0.467568	0.442251	3.609394	0.571197	1.480831
		33.282392	0.171397	0.989489	5.769089	0.986868	5.654795
		-	-	0.296470	-	-	-
100.0	2.8424	16.268599	0.467763	0.439578	4.033435	0.569994	1.484280
		41.770317	0.172925	0.989253	6.462996	0.986703	5.638939
		-	-	0.295027	-	-	-

ELLIPTIC LOW-PASS FILTER DATA

N= 5   PRW= 2.0 DB

MSL	TW	A	B	C	WZ	WM	KM
30.0	0.0659	1.913632 1.173393 -	0.294821 0.044073 -	0.676164 0.988405 0.357006	1.383341 1.083233 -	0.767005 0.988491 -	1.930311 3.673729 -
35.0	0.1098	1.284322 2.270890 -	0.314214 0.057182 -	0.619156 0.983731 0.324878	1.133279 1.506947 -	0.698319 0.989749 -	1.501360 9.857439 -
40.0	0.1681	1.435744 2.722833 -	0.327127 0.069466 -	0.572573 0.979131 0.300797	1.198225 1.650101 -	0.674796 0.986923 -	1.564307 9.147511 -
45.0	0.2419	1.636338 3.293356 -	0.335666 0.080502 -	0.535178 0.974850 0.282514	1.279194 1.814761 -	0.655354 0.984324 -	1.617319 8.658723 -
50.0	0.3324	1.897108 4.012746 -	0.341310 0.090121 -	0.505442 0.971019 0.268499	1.377355 2.003184 -	0.639474 0.982025 -	1.661424 8.311556 -
55.0	0.4408	2.232109 4.919259 -	0.345054 0.098323 -	0.481912 0.967686 0.257674	1.494024 2.217940 -	0.626605 0.980048 -	1.697773 8.059078 -
60.0	0.5686	2.659264 6.061143 -	0.347554 0.105202 -	0.463330 0.964846 0.249265	1.630725 2.461939 -	0.616235 0.978379 -	1.727508 7.872069 -
65.0	0.7174	3.201365 7.499194 -	0.349239 0.110901 -	0.448666 0.962466 0.242703	1.789236 2.738466 -	0.607910 0.976992 -	1.751688 7.731548 -
70.0	0.8891	3.887302 9.309986 -	0.350386 0.115578 -	0.437090 0.960493 0.237564	1.971624 3.051227 -	0.601246 0.975850 -	1.771259 7.624755 -
75.0	1.0859	4.753614 11.589945 -	0.351177 0.119389 -	0.427947 0.958874 0.233527	2.180278 3.404401 -	0.595921 0.974919 -	1.787038 7.542859 -
80.0	1.3106	5.846441 14.460486 -	0.351729 0.122478 -	0.420721 0.957554 0.230349	2.417941 3.802695 -	0.591673 0.974163 -	1.799723 7.479605 -
85.0	1.5660	7.223982 18.074475 -	0.352120 0.124970 -	0.415005 0.956483 0.227842	2.687747 4.251409 -	0.588287 0.973553 -	1.809895 7.430470 -
90.0	1.8555	8.959599 22.624369 -	0.352400 0.126975 -	0.410481 0.955620 0.225862	2.993259 4.756508 -	0.585590 0.973062 -	1.818036 7.392128 -
95.0	2.1829	11.145721 28.352470 -	0.352603 0.128582 -	0.406897 0.954925 0.224297	3.338521 5.324704 -	0.583444 0.972669 -	1.824543 7.362099 -
100.0	2.5525	13.898769 35.563824 -	0.352753 0.129869 -	0.404058 0.954367 0.223057	3.728105 5.963541 -	0.581737 0.972353 -	1.829735 7.338513 -

APPENDIX C

N= 5   PRW= 3.0 DB

MSL	TW	A	B	C	WZ	WM	KM
30.0	0.0502	1.775450 1.133895 -	0.233483 0.031739 -	0.677491 0.984644 0.301724	1.332460 1.064845 -	0.786175 0.988696 -	2.289057 4.219991 -
35.0	0.0878	2.095481 1.228550 -	0.251399 0.042539 -	0.617274 0.978423 0.272877	1.447578 1.108400 -	0.748662 0.985123 -	2.300037 4.818638 -
40.0	0.1393	2.501033 1.360300 -	0.263363 0.052862 -	0.567823 0.972271 0.251277	1.581465 1.166319 -	0.716757 0.981785 -	2.297723 5.389557 -
45.0	0.2058	3.013444 1.536974 -	0.271273 0.062262 -	0.528030 0.966529 0.234894	1.735927 1.239747 -	0.690193 0.978801 -	2.288855 5.916865 -
50.0	0.2885	3.659862 1.768413 -	0.276486 0.070534 -	0.496362 0.961381 0.222346	1.913077 1.329817 -	0.668390 0.976214 -	2.277416 6.390774 -
55.0	0.3885	2.067167 4.474636 -	0.279929 0.077634 -	0.471303 0.956899 0.212663	1.437764 2.115333 -	0.640683 0.975833 -	1.994487 9.924302 -
60.0	0.5071	2.449260 5.501115 -	0.282214 0.083619 -	0.451524 0.953080 0.205147	1.565011 2.345446 -	0.628356 0.973716 -	2.034325 9.670144 -
65.0	0.6459	2.935099 6.793943 -	0.283743 0.088596 -	0.435925 0.949877 0.199286	1.713213 2.606519 -	0.618455 0.971946 -	2.066726 9.479856 -
70.0	0.8068	3.550586 8.421957 -	0.284776 0.092691 -	0.423619 0.947223 0.194697	1.884300 2.902061 -	0.610528 0.970485 -	2.092953 9.335645 -
75.0	0.9918	4.328514 10.471850 -	0.285481 0.096035 -	0.413907 0.945044 0.191095	2.080508 3.236024 -	0.604192 0.969289 -	2.114104 9.225300 -
80.0	1.2032	5.310316 13.052783 -	0.285969 0.098750 -	0.406235 0.943269 0.188260	2.304412 3.612864 -	0.599137 0.968316 -	2.131110 9.140221 -
85.0	1.4440	6.548281 16.302199 -	0.286311 0.100942 -	0.400170 0.941829 0.186025	2.558961 4.037598 -	0.595108 0.967529 -	2.144748 9.074222 -
90.0	1.7173	8.108337 20.393140 -	0.286553 0.102707 -	0.395371 0.940667 0.184260	2.847514 4.515876 -	0.591900 0.966894 -	2.155665 9.022776 -
95.0	2.0267	10.073564 25.543464 -	0.286728 0.104124 -	0.391572 0.939733 0.182865	3.173888 5.054054 -	0.589346 0.966385 -	2.164391 8.982520 -
100.0	2.3763	12.548619 32.027449 -	0.286855 0.105259 -	0.388562 0.938983 0.181761	3.542403 5.659280 -	0.587314 0.965976 -	2.171355 8.950923 -

ELLIPTIC LOW-PASS FILTER DATA

N= 6   PRW= 0.1 DB

MSL	TW	A	B	C	WZ	WM	KM
30.0	0.1025	1.249523	1.195815	0.621688	1.117821	–	–
		8.935115	0.391996	0.941234	2.989166	0.920443	2.285018
		1.727485	0.075069	1.054358	1.314338	1.021157	5.390384
35.0	0.1483	1.362594	1.140090	0.540917	1.167302	–	–
		10.844578	0.436713	0.908030	3.293111	0.892364	2.075348
		1.966549	0.093122	1.064273	1.402337	1.024587	5.141050
40.0	0.2045	1.506893	1.091960	0.480628	1.227556	–	–
		2.259755	0.473312	0.877158	1.503248	0.801290	1.407242
		13.137107	0.110432	1.073388	3.624515	1.032574	8.632161
45.0	0.2720	1.688022	1.051294	0.434969	1.299239	–	–
		2.617953	0.502928	0.849449	1.618009	0.786154	1.416465
		15.897042	0.126535	1.081584	3.987109	1.035575	7.678457
50.0	0.3513	1.912825	1.017329	0.399933	1.383049	–	–
		3.054375	0.526750	0.825165	1.747677	0.773221	1.422837
		19.226054	0.141167	1.088829	4.384752	1.038111	6.993207
55.0	0.4433	2.189692	0.989137	0.372737	1.479761	–	–
		3.585140	0.545862	0.804229	1.893447	0.762256	1.427220
		23.246915	0.154218	1.095147	4.821506	1.040234	6.487393
60.0	0.5487	2.528898	0.965820	0.351416	1.590251	–	–
		4.229849	0.561193	0.786387	2.056660	0.753012	1.430226
		28.108017	0.165691	1.100598	5.301699	1.041998	6.106001
65.0	0.6685	2.942999	0.946569	0.334559	1.715517	–	–
		5.012310	0.573506	0.771304	2.238819	0.745254	1.432282
		33.988821	0.175658	1.105262	5.829993	1.043458	5.813509
70.0	0.8038	3.447303	0.930691	0.321137	1.856692	–	–
		5.961421	0.583415	0.758630	2.441602	0.738766	1.433688
		41.106481	0.184237	1.109225	6.411434	1.044663	5.586126
75.0	0.9558	4.060446	0.917602	0.310384	2.015055	–	–
		7.112232	0.591410	0.748025	2.666877	0.733354	1.434647
		49.723842	0.191564	1.112574	7.051513	1.045655	5.407411
80.0	1.1258	4.805077	0.906813	0.301726	2.192049	–	–
		8.507247	0.597878	0.739180	2.916719	0.728850	1.435301
		60.159154	0.197783	1.115393	7.756233	1.046471	5.265693
85.0	1.3156	5.708697	0.897921	0.294726	2.389288	–	–
		10.197983	0.603122	0.731820	3.193428	0.725106	1.435748
		72.797788	0.203036	1.117756	8.532162	1.047143	5.152499
90.0	1.5267	6.804685	0.890590	0.289046	2.608579	–	–
		12.246889	0.607386	0.725706	3.499556	0.721999	1.436052
		88.106554	0.207455	1.119732	9.386509	1.047695	5.061554
95.0	1.7611	8.133511	0.884547	0.284423	2.851931	–	–
		14.729620	0.610859	0.720634	3.837919	0.719423	1.436259
		106.650727	0.211159	1.121381	10.327184	1.048150	4.988134
100.0	2.0210	9.744291	0.879565	0.280653	3.121585	–	–
		17.737943	0.613695	0.716431	4.211644	0.717289	1.436401
		129.115735	0.214256	1.122753	11.362910	1.048524	4.928625

APPENDIX C

N= 6   PRW= 0.5 DB

MSL	TW	A	B	C	WZ	WM	KM
30.0	0.0539	1.132707	0.850749	0.423655	1.064287	–	–
		6.726475	0.243303	0.871392	2.593545	0.912785	3.387914
		1.461602	0.040219	1.010435	1.208967	1.003001	7.761155
35.0	0.0857	1.208684	0.807840	0.362848	1.099402	–	–
		8.202218	0.277497	0.832484	2.863951	0.886300	3.006363
		1.637636	0.052686	1.012420	1.279702	1.003270	7.335996
40.0	0.1270	1.309514	0.770074	0.317565	1.144340	–	–
		9.965458	0.305751	0.796492	3.156811	0.861341	2.740985
		1.855733	0.065032	1.014167	1.362253	1.003481	7.065146
45.0	0.1785	1.439472	0.737775	0.283385	1.199780	–	–
		12.080130	0.328726	0.764340	3.475648	0.838631	2.549408
		2.123938	0.076782	1.015670	1.457374	1.003664	6.888724
50.0	0.2410	1.603682	0.710575	0.257260	1.266366	–	–
		14.623574	0.347239	0.736289	3.824078	0.818472	2.407205
		2.452144	0.087635	1.016942	1.565932	1.003835	6.772113
55.0	0.3151	1.808399	0.687870	0.237064	1.344767	–	–
		2.852463	0.362090	0.712210	1.688924	0.778634	1.862111
		17.689265	0.097433	1.018008	4.205861	1.006321	9.773855
60.0	0.4014	2.061294	0.669013	0.221292	1.435721	–	–
		3.339673	0.373987	0.691772	1.827477	0.766837	1.868922
		21.390154	0.106123	1.018893	4.624949	1.006330	9.073738
65.0	0.5008	2.371769	0.653399	0.208869	1.540055	–	–
		3.931763	0.383522	0.674559	1.982867	0.756868	1.873590
		25.862764	0.113723	1.019625	5.085545	1.006295	8.544953
70.0	0.6141	2.751325	0.640493	0.199010	1.658712	–	–
		4.650598	0.391179	0.660141	2.156525	0.748487	1.876783
		31.272187	0.120298	1.020228	5.592154	1.006234	8.138873
75.0	0.7425	3.214001	0.629836	0.191135	1.792764	–	–
		5.522725	0.397340	0.648112	2.350048	0.741467	1.878964
		37.818150	0.125936	1.020724	6.149646	1.006159	7.822831
80.0	0.8870	3.776896	0.621042	0.184812	1.943424	–	–
		6.580351	0.402313	0.638104	2.565219	0.735606	1.880453
		45.742379	0.130736	1.021132	6.763311	1.006080	7.574203
85.0	1.0490	4.460804	0.613787	0.179710	2.112062	–	–
		7.862536	0.406336	0.629793	2.804021	0.730723	1.881468
		55.337514	0.134800	1.021467	7.438919	1.006003	7.376894
90.0	1.2299	5.290987	0.607803	0.175579	2.300215	–	–
		9.416634	0.409600	0.622903	3.068654	0.726662	1.882161
		66.957918	0.138224	1.021742	8.182782	1.005930	7.219199
95.0	1.4314	6.298108	0.602867	0.172223	2.509603	–	–
		11.300041	0.412254	0.617195	3.361553	0.723290	1.882633
		81.032746	0.141100	1.021969	9.001819	1.005863	7.092433
100.0	1.6553	7.519366	0.598795	0.169488	2.742146	–	–
		13.582319	0.414416	0.612472	3.685420	0.720493	1.882955
		98.081841	0.143506	1.022155	9.903628	1.005804	6.990049

ELLIPTIC LOW-PASS FILTER DATA

N= 6   PRW= 1.0 DB

MSL	TW	A	B	C	WZ	WM	KM
30.0	0.0380	1.094749	0.699867	0.363422	1.046303	-	-
		5.892956	0.183080	0.859924	2.427541	0.915157	4.364133
		1.366253	0.027615	1.000709	1.168868	0.999121	9.730561
35.0	0.0639	1.156482	0.663497	0.308546	1.075398	0.081353	1.000214
		7.209219	0.212409	0.818473	2.684999	0.888952	3.817049
		1.518324	0.037528	1.000155	1.232203	0.998367	9.132385
40.0	0.0989	1.240615	0.631034	0.267756	1.113829	0.107246	1.000881
		8.777806	0.236837	0.779910	2.962736	0.864009	3.441660
		1.708092	0.047572	0.999426	1.306940	0.997552	8.754708
45.0	0.1436	1.351011	0.603032	0.237044	1.162330	0.119204	1.001746
		10.654846	0.256790	0.745345	3.264176	0.841154	3.173293
		1.942541	0.057288	0.998598	1.393751	0.996742	8.510148
50.0	0.1989	1.492203	0.579320	0.213632	1.221558	0.125395	1.002664
		12.908535	0.272904	0.715137	3.592845	0.820766	2.975507
		2.230307	0.066367	0.997733	1.493421	0.995980	8.349130
55.0	0.2653	1.669674	0.559453	0.195582	1.292159	0.128709	1.003557
		15.621428	0.285839	0.689186	3.952395	0.802923	2.826333
		2.582002	0.074635	0.996880	1.606861	0.995290	8.241935
60.0	0.3435	1.890135	0.542912	0.181525	1.374822	0.130471	1.004385
		18.893319	0.296199	0.667156	4.346645	0.787522	2.711797
		3.010604	0.082014	0.996072	1.735109	0.994683	8.170038
65.0	0.4343	2.161819	0.529190	0.170479	1.470312	0.131367	1.005131
		22.844814	0.304495	0.648608	4.779625	0.774365	2.622619
		3.531936	0.088500	0.995329	1.879345	0.994158	8.121573
70.0	0.5384	2.494809	0.517835	0.161733	1.579497	0.131771	1.005789
		27.621718	0.311149	0.633078	5.255637	0.763211	2.552412
		4.165249	0.094131	0.994659	2.040894	0.993710	8.088791
75.0	0.6569	2.901434	0.508451	0.154760	1.703360	0.131899	1.006361
		4.933929	0.316497	0.620128	2.221245	0.745911	2.249993
		33.400375	0.098973	0.994067	5.779306	0.994419	9.787492
80.0	0.7907	3.396727	0.500701	0.149171	1.843021	0.131876	1.006853
		5.866364	0.320807	0.609360	2.422058	0.739315	2.252284
		40.394159	0.103105	0.993550	6.355640	0.993965	9.443643
85.0	0.9411	3.998989	0.494305	0.144669	1.999747	0.131773	1.007273
		6.996989	0.324290	0.600423	2.645182	0.733812	2.253847
		48.861338	0.106610	0.993103	6.990089	0.993572	9.171847
90.0	1.1094	4.730468	0.489027	0.141027	2.174964	0.131635	1.007628
		8.367561	0.327112	0.593017	2.892674	0.729230	2.254913
		59.114609	0.109567	0.992721	7.688603	0.993234	8.955317
95.0	1.2973	5.618183	0.484672	0.138072	2.370271	0.131486	1.007928
		10.028697	0.329404	0.586886	3.166812	0.725422	2.255640
		71.532638	0.112053	0.992395	8.457697	0.992946	8.781711
100.0	1.5063	6.694925	0.481080	0.135667	2.587455	0.131341	1.008180
		12.041756	0.331270	0.581814	3.470123	0.722261	2.256135
		86.574113	0.114135	0.992120	9.304521	0.992701	8.641796

APPENDIX C

N= 6   PRW= 2.0 DB

MSL	TW	A	B	C	WZ	WM	KM
30.0	0.0246	1.278419	0.546002	0.321085	1.130672	0.324948	1.055278
		5.089564	0.126088	0.861163	2.256006	0.921953	6.142542
		1.062730	0.016829	0.995570	1.030888	0.995610	3.870010
35.0	0.0447	1.407193	0.517013	0.269645	1.186251	0.305525	1.063559
		6.254789	0.150114	0.817265	2.500958	0.895902	5.266015
		1.110700	0.024100	0.993135	1.053897	0.993957	4.478264
40.0	0.0732	1.569612	0.490548	0.231526	1.252842	0.287979	1.069574
		7.639336	0.170388	0.775852	2.763935	0.870685	4.677360
		1.178595	0.031719	0.990452	1.085631	0.992304	5.092437
45.0	0.1109	1.771634	0.467420	0.202930	1.331027	0.272724	1.073848
		9.291646	0.187077	0.738399	3.048220	0.847302	4.263007
		1.269973	0.039267	0.987696	1.126931	0.990730	5.695349
50.0	0.1587	2.020682	0.447677	0.181217	1.421507	0.259735	1.076843
		11.271059	0.200611	0.705481	3.357240	0.826266	3.961110
		1.388860	0.046443	0.985003	1.178499	0.989283	6.272480
55.0	0.2171	2.325932	0.431048	0.164541	1.525101	0.248811	1.078922
		13.649666	0.211496	0.677105	3.694545	0.807743	3.735366
		1.540035	0.053060	0.982467	1.240981	0.987985	6.812791
60.0	0.2869	2.698639	0.417155	0.151603	1.642754	0.239692	1.080357
		16.514719	0.220216	0.652968	4.063831	0.791685	3.563176
		1.729312	0.059021	0.980144	1.315033	0.986843	7.308918
65.0	0.3687	3.152560	0.405603	0.141471	1.775545	0.232115	1.081342
		19.971682	0.227195	0.632623	4.468969	0.777922	3.429785
		1.963813	0.064298	0.978061	1.401361	0.985852	7.756900
70.0	0.4633	3.704452	0.396027	0.133473	1.924695	0.225838	1.082018
		24.148031	0.232786	0.615581	4.914065	0.766226	3.325187
		2.252275	0.068905	0.976222	1.500758	0.985001	8.155634
75.0	0.5715	4.374695	0.388105	0.127115	2.091577	0.220647	1.082480
		29.197893	0.237273	0.601368	5.403507	0.756349	3.242368
		2.605395	0.072883	0.974621	1.614124	0.984276	8.506209
80.0	0.6943	5.188041	0.381558	0.122030	2.277727	0.216359	1.082795
		35.307703	0.240883	0.589550	5.942029	0.748049	3.176285
		3.036240	0.076289	0.973241	1.742481	0.983663	8.811253
85.0	0.8328	6.174525	0.376151	0.117943	2.484859	0.212819	1.083010
		42.703074	0.243796	0.579744	6.534759	0.741100	3.123227
		3.560734	0.079185	0.972060	1.886991	0.983147	9.074362
90.0	0.9883	7.370582	0.371687	0.114643	2.714881	0.209897	1.083157
		51.657107	0.246153	0.571620	7.187288	0.735299	3.080412
		4.198257	0.081634	0.971057	2.048965	0.982713	9.299632
95.0	1.1622	8.820382	0.368004	0.111969	2.969913	0.207487	1.083258
		62.500488	0.248064	0.564895	7.905725	0.730467	3.045720
		4.972358	0.083695	0.970209	2.229879	0.982351	9.491317
100.0	1.3561	10.577472	0.364964	0.109796	3.252303	0.205499	1.083326
		75.633682	0.249619	0.559334	8.696763	0.726449	3.017518
		5.911629	0.085425	0.969496	2.431384	0.982049	9.653582

ELLIPTIC LOW-PASS FILTER DATA

N= 6   PRW= 3.0 DB

MSL	TW	A	B	C	WZ	WM	KM
30.0	0.0178	1.229027	0.454392	0.306861	1.108615	0.390166	1.142843
		4.615053	0.094841	0.868790	2.148267	0.928555	7.999986
		1.046388	0.011509	0.994431	1.022931	0.995908	4.411593
35.0	0.0344	1.343891	0.430195	0.255655	1.159263	0.360974	1.156824
		5.692316	0.115471	0.823674	2.385858	0.902642	6.746799
		1.086290	0.017212	0.991265	1.042252	0.993998	5.149504
40.0	0.0589	1.490100	0.407624	0.217811	1.220697	0.336338	1.166825
		6.970040	0.133089	0.780489	2.640083	0.877163	5.921037
		1.144555	0.023362	0.987735	1.069839	0.991983	5.902675
45.0	0.0923	1.673003	0.387667	0.189508	1.293446	0.315778	1.173864
		8.491988	0.147699	0.741055	2.914102	0.853276	5.347676
		1.224624	0.029577	0.984078	1.106627	0.989984	6.649512
50.0	0.1353	1.899308	0.370511	0.168086	1.378154	0.298737	1.178766
		10.312213	0.159597	0.706177	3.211263	0.831618	4.934119
		1.330266	0.035570	0.980485	1.153372	0.988086	7.370982
55.0	0.1888	2.177344	0.355999	0.151686	1.475583	0.284672	1.182155
		12.496635	0.169186	0.675988	3.535058	0.812440	4.627218
		1.465879	0.041156	0.977089	1.210735	0.986338	8.051955
60.0	0.2533	2.517356	0.343839	0.138999	1.586618	0.273090	1.184487
		15.125158	0.176874	0.650241	3.889108	0.795746	4.394477
		1.636764	0.046227	0.973969	1.279361	0.984768	8.681738
65.0	0.3294	2.931891	0.333709	0.129091	1.712277	0.263564	1.186086
		18.294412	0.183026	0.628504	4.277197	0.781396	4.214988
		1.849403	0.050743	0.971167	1.359928	0.983381	9.253956
70.0	0.4180	3.436251	0.325302	0.121289	1.853713	0.255732	1.187181
		22.121184	0.187952	0.610277	4.703316	0.769713	4.074734
		2.111749	0.054704	0.968691	1.453186	0.982173	9.766004
75.0	0.5198	4.049057	0.318340	0.115101	2.012227	0.249295	1.187929
		26.746659	0.191901	0.595068	5.171717	0.758835	3.963992
		2.433545	0.058136	0.966532	1.559982	0.981132	10.218270
80.0	0.6357	4.792940	0.312583	0.110162	2.189278	0.244003	1.188440
		32.341579	0.195075	0.582417	5.686966	0.750136	3.875823
		2.826709	0.061082	0.964670	1.681282	0.980243	10.613330
85.0	0.7668	5.695370	0.307827	0.106198	2.386497	0.239650	1.188789
		39.112529	0.197633	0.571919	6.254001	0.742846	3.805157
		3.305780	0.063593	0.963078	1.818180	0.979488	10.955199
90.0	0.9143	6.789677	0.303900	0.103002	2.605701	0.236069	1.189026
		47.309545	0.199701	0.563220	6.878193	0.736755	3.748214
		3.888459	0.065720	0.961724	1.971918	0.978850	11.248712
95.0	1.0796	8.116273	0.300658	0.100416	2.848907	0.233121	1.189188
		57.235343	0.201376	0.556021	7.565404	0.731679	3.702128
		4.596273	0.067513	0.960579	2.143892	0.978314	11.499043
100.0	1.2640	9.724154	0.297983	0.098316	3.118358	0.230694	1.189299
		69.256502	0.202737	0.550067	8.322049	0.727456	3.664696
		5.455367	0.069019	0.959616	2.335673	0.977865	11.711364

APPENDIX C

N= 7   PRW= 0.1 DB

MSL	TW	A	B	C	WZ	WM	KM
30.0	0.0479	1.112775	0.722673	0.764652	1.054882	–	–
		2.677955	0.192302	0.970795	1.636446	0.965245	3.337216
		1.298987	0.035531	1.025709	1.139731	1.010125	6.063045
		–	–	0.704469	–	–	–
35.0	0.0728	1.170281	0.739095	0.692282	1.081795	0.122494	1.000153
		3.094310	0.228919	0.951081	1.759065	0.949853	3.023270
		1.408614	0.046632	1.032145	1.186850	1.012481	5.886876
		–	–	0.642392	–	–	–
40.0	0.1045	1.244729	0.745651	0.631257	1.115674	0.215049	1.002002
		3.581852	0.262332	0.930725	1.892578	0.934295	2.795239
		1.542665	0.058023	1.038486	1.242041	1.014838	5.800668
		–	–	0.594039	–	–	–
45.0	0.1434	1.338586	0.746175	0.580438	1.156973	0.255020	1.005148
		4.153513	0.292130	0.910729	2.038017	0.919216	2.624018
		1.704863	0.069297	1.044556	1.305704	1.017150	5.768985
		–	–	0.555714	–	–	–
50.0	0.1901	1.454690	0.743211	0.538372	1.206105	0.277094	1.008919
		4.824587	0.318270	0.891769	2.196494	0.905035	2.492213
		1.899657	0.080141	1.050240	1.378280	1.019377	5.770431
		–	–	0.524936	–	–	–
55.0	0.2449	1.596371	0.738398	0.503628	1.263476	0.290396	1.012904
		2.132355	0.340930	0.874245	1.460258	0.861099	1.769861
		5.613135	0.090343	1.055469	2.369205	1.024453	9.252640
		–	–	0.499970	–	–	–
60.0	0.3082	1.767587	0.732760	0.474931	1.329506	0.298801	1.016853
		2.409283	0.360412	0.858353	1.552187	0.852651	1.793380
		6.540448	0.099769	1.060212	2.557430	1.026312	8.667041
		–	–	0.479559	–	–	–
65.0	0.3805	1.973054	0.726924	0.451198	1.404654	0.304274	1.020616
		2.737960	0.377065	0.844144	1.654678	0.845308	1.813092
		7.631615	0.108356	1.064464	2.762538	1.027958	8.213440
		–	–	0.462766	–	–	–
70.0	0.4622	2.218385	0.721256	0.431533	1.489424	0.307912	1.024109
		3.127303	0.391243	0.831579	1.768418	0.838961	1.829631
		8.916185	0.116084	1.068242	2.985998	1.029406	7.856825
		–	–	0.448880	–	–	–
75.0	0.5539	2.510260	0.715956	0.415201	1.584380	0.310366	1.027293
		3.587870	0.403282	0.820559	1.894167	0.833495	1.843525
		10.428963	0.122974	1.071573	3.229391	1.030673	7.573004
		–	–	0.437349	–	–	–
80.0	0.6563	2.856610	0.711122	0.401607	1.690151	0.312040	1.030154
		4.132150	0.413487	0.810956	2.032769	0.828804	1.855212
		12.210953	0.129068	1.074491	3.494417	1.031776	7.344808
		–	–	0.427741	–	–	–
85.0	0.7700	3.266834	0.706787	0.390267	1.807439	0.313192	1.032699
		4.774902	0.422129	0.802632	2.185155	0.824787	1.865053
		14.310454	0.134421	1.077033	3.782916	1.032732	7.159777
		–	–	0.419713	–	–	–
90.0	0.8957	3.752062	0.702947	0.380788	1.937024	0.313989	1.034944
		5.533553	0.429444	0.795444	2.352350	0.821355	1.873348
		16.784392	0.139098	1.079240	4.096876	1.033559	7.008678
		–	–	0.412989	–	–	–

ELLIPTIC LOW-PASS FILTER DATA

N= 7   PRW= 0.1 DB (CONTINUED)

MSL	TW	A	B	C	WZ	WM	KM
95.0	1.0342	4.325451	0.699576	0.372849	2.079772	0.314543	1.036911
		6.428667	0.435633	0.789258	2.535482	0.818425	1.880346
		19.699819	0.143165	1.081148	4.438448	1.034271	6.884555
		-	-	0.407346	-	-	-
100.0	1.1866	5.002564	0.696637	0.366188	2.236641	0.314930	1.038625
		7.484539	0.440869	0.783947	2.735789	0.815928	1.886256
		23.135859	0.146689	1.082793	4.809975	1.034883	6.782079
		-	-	0.402601	-	-	-

N= 7   PRW= 0.5 DB

MSL	TW	A	B	C	WZ	WM	KM
30.0	0.0230	1.055400	0.489793	0.665498	1.027327	0.510524	1.053154
		2.183446	0.109644	0.941945	1.477649	0.962749	5.083844
		1.178193	0.017353	1.004500	1.085446	1.001304	8.564894
		-	-	0.517530	-	-	-
35.0	0.0391	1.092378	0.505408	0.596678	1.045169	0.500644	1.072538
		2.515426	0.135994	0.917677	1.586010	0.947587	4.526273
		1.257910	0.024482	1.005766	1.121566	1.001538	8.260385
		-	-	0.468549	-	-	-
40.0	0.0611	1.143142	0.512467	0.538491	1.069178	0.488315	1.090774
		2.903858	0.160616	0.892646	1.704071	0.931908	4.126614
		1.357785	0.032148	1.006991	1.165240	1.001755	8.111420
		-	-	0.430200	-	-	-
45.0	0.0397	1.209863	0.514138	0.490004	1.099938	0.475727	1.107559
		3.358629	0.182936	0.868078	1.832656	0.916444	3.829272
		1.480633	0.039997	1.008139	1.216833	1.001965	8.055488
		-	-	0.399685	-	-	-
50.0	0.1254	1.294882	0.512588	0.449901	1.137929	0.463823	1.122785
		3.891641	0.202737	0.844810	1.972724	0.901708	3.601880
		1.630030	0.047739	1.009189	1.276726	1.002174	8.055712
		-	-	0.375107	-	-	-
55.0	0.1687	1.400852	0.509251	0.416834	1.183576	0.452988	1.136454
		4.517069	0.220035	0.823342	2.125340	0.888015	3.424332
		1.809930	0.055160	1.010131	1.345336	1.002386	8.089247
		-	-	0.355126	-	-	-
60.0	0.2198	1.530872	0.505044	0.389584	1.237284	0.443341	1.148622
		5.251693	0.234983	0.803914	2.291657	0.875532	3.283476
		2.025294	0.062116	1.010966	1.423128	1.002598	8.141516
		-	-	0.338765	-	-	-
65.0	0.2793	1.688607	0.500537	0.367106	1.299464	0.434870	1.159379
		6.115319	0.247802	0.786583	2.472917	0.864317	3.170332
		2.281981	0.068521	1.011698	1.510623	1.002807	8.203095
		-	-	0.325289	-	-	-
70.0	0.3476	1.878424	0.496073	0.348529	1.370556	0.427499	1.168833
		7.131294	0.258740	0.771291	2.670448	0.854352	3.078554
		2.586961	0.074336	1.012336	1.608403	1.003011	8.267902
		-	-	0.314136	-	-	-
75.0	0.4252	2.105523	0.491847	0.333141	1.451042	0.421125	1.177099
		8.327121	0.268039	0.757910	2.885675	0.845575	3.003521
		2.948511	0.079554	1.012887	1.717123	1.003205	8.332076
		-	-	0.304871	-	-	-

APPENDIX C

N= 7   PRW= 0.5 DB (CONTINUED)

MSL	TW	A	B	C	WZ	WM	KM
80.0	0.5125	2.376095	0.487959	0.320365	1.541459	0.415639	1.184296
		9.735195	0.275927	0.746276	3.120127	0.837899	2.941791
		3.376437	0.084194	1.013361	1.837508	1.003388	8.393256
		-	-	0.297148	-	-	-
85.0	0.6101	2.697498	0.484451	0.309732	1.642406	0.410931	1.190538
		11.393673	0.282608	0.736209	3.375452	0.831221	2.890743
		3.882343	0.088287	1.013768	1.970366	1.003556	8.450097
		-	-	0.290693	-	-	-
90.0	0.7187	3.078458	0.481330	0.300861	1.754554	0.406902	1.195935
		13.347513	0.288263	0.727532	3.653425	0.825437	2.848351
		4.479947	0.091875	1.014117	2.116589	1.003710	8.501934
		-	-	0.285286	-	-	-
95.0	0.8390	3.529312	0.478581	0.293446	1.878646	0.403460	1.200589
		5.185452	0.293046	0.720076	2.277159	0.814758	2.559723
		15.649677	0.095003	1.014415	3.955967	1.004628	9.928650
		-	-	0.280748	-	-	-
100.0	0.9718	4.062289	0.476178	0.287235	2.015512	0.400523	1.204591
		6.017989	0.297093	0.713685	2.453159	0.811258	2.569918
		18.362586	0.097720	1.014669	4.285159	1.004658	9.752821
		-	-	0.276933	-	-	-

N= 7   PRW= 1.0 DB

MSL	TW	A	B	C	WZ	WM	KM
30.0	0.0154	1.135869	0.388934	0.644593	1.065772	0.638005	1.204021
		1.993333	0.078251	0.940082	1.411855	0.965184	6.588364
		1.037690	0.011284	1.000289	1.018671	0.998412	3.342125
		-	-	0.433912	-	-	-
35.0	0.0279	1.203716	0.404339	0.575473	1.097140	0.608789	1.243691
		2.292727	0.099804	0.914581	1.514175	0.950280	5.802218
		1.066795	0.016689	1.000067	1.032858	0.997879	3.871185
		-	-	0.391155	-	-	-
40.0	0.0460	1.108303	0.411828	0.516833	1.052760	0.559508	1.205598
		2.643201	0.120286	0.887946	1.625792	0.934587	5.245106
		1.290109	0.022689	0.999722	1.135830	0.998846	9.958737
		-	-	0.357648	-	-	-
45.0	0.0702	1.164343	0.414231	0.467886	1.079047	0.537467	1.231465
		3.053379	0.139072	0.861574	1.747392	0.918900	4.833973
		1.397605	0.028973	0.999281	1.182203	0.998377	9.870596
		-	-	0.330964	-	-	-
50.0	0.1013	1.237124	0.413531	0.427381	1.112261	0.518079	1.254420
		3.533796	0.155873	0.836446	1.879839	0.903799	4.521420
		1.529246	0.035277	0.998776	1.236627	0.997905	9.861803
		-	-	0.309462	-	-	-
55.0	0.1395	1.329082	0.411059	0.393993	1.152858	0.501213	1.274699
		4.097098	0.170632	0.813166	2.024129	0.889660	4.278462
		1.688676	0.041396	0.998238	1.299491	0.997451	9.900218
		-	-	0.291977	-	-	-
60.0	0.1859	1.443008	0.407678	0.366501	1.201253	0.486649	1.292535
		4.758317	0.183436	0.792040	2.181357	0.876696	4.086366
		1.880263	0.047186	0.997693	1.371227	0.997029	9.965479
		-	-	0.277657	-	-	-

ELLIPTIC LOW-PASS FILTER DATA

N= 7   PRW= 1.0 DB (CONTINUED)

MSL	TW	A	B	C	WZ	WM	KM
65.0	0.2395	2.109229	0.403934	0.343848	1.452319	0.485278	1.363808
		5.535229	0.194444	0.773161	2.352707	0.864995	3.932472
		1.582184	0.052558	0.997158	1.257849	0.995532	7.072458
		–	–	0.265862	–	–	–
70.0	0.3020	2.381798	0.400158	0.325150	1.543308	0.472991	1.371961
		6.448804	0.203853	0.756485	2.539450	0.854564	3.807898
		1.750501	0.057462	0.996650	1.323065	0.995312	7.523054
		–	–	0.256102	–	–	–
75.0	0.3734	2.705370	0.396543	0.309684	1.644801	0.462469	1.378406
		7.523752	0.211860	0.741883	2.742946	0.845352	3.706225
		1.952600	0.061882	0.996175	1.397355	0.995130	7.936231
		–	–	0.247994	–	–	–
80.0	0.4543	3.088724	0.393193	0.296858	1.757477	0.453480	1.383519
		8.789182	0.218656	0.729181	2.964656	0.837278	3.622690
		2.194007	0.065827	0.995740	1.481218	0.994979	8.311023
		–	–	0.241237	–	–	–
85.0	0.5451	3.542256	0.390155	0.286196	1.882088	0.445814	1.387593
		10.279379	0.224415	0.718191	3.206147	0.830243	3.553687
		2.481294	0.069317	0.995346	1.575212	0.994854	8.647805
		–	–	0.235590	–	–	–
90.0	0.6464	4.078264	0.387442	0.277313	2.019471	0.439284	1.390853
		12.034731	0.229290	0.708717	3.469111	0.824141	3.496436
		2.822270	0.072383	0.994993	1.679961	0.994750	8.947964
		–	–	0.230861	–	–	–
95.0	0.7590	4.711283	0.385045	0.269894	2.170549	0.433728	1.393474
		14.102821	0.233414	0.700577	3.755372	0.818871	3.448764
		3.226188	0.075062	0.994680	1.796159	0.994665	9.213590
		–	–	0.226892	–	–	–
100.0	0.8836	5.458479	0.382945	0.263686	2.336339	0.429003	1.395593
		16.539723	0.236902	0.693600	4.066906	0.814332	3.408948
		3.704009	0.077392	0.994405	1.924580	0.994594	9.447224
		–	–	0.223557	–	–	–

N= 7   PRW= 2.0 DB

MSL	TW	A	B	C	WZ	WM	KM
35.0	0.0185	1.153933	0.302998	0.570029	1.074213	0.666616	1.473498
		2.076068	0.066688	0.918759	1.440857	0.955519	8.044843
		1.045047	0.010134	0.997113	1.022275	0.997461	4.624116
		–	–	0.311397	–	–	–
40.0	0.0326	1.226945	0.310940	0.509326	1.107675	0.632254	1.525676
		2.389539	0.082728	0.891018	1.545814	0.939970	7.188403
		1.077551	0.014453	0.995648	1.038051	0.996498	5.334952
		–	–	0.283165	–	–	–
45.0	0.0524	1.319215	0.314164	0.458470	1.148571	0.601728	1.567573
		2.756462	0.097697	0.863070	1.660260	0.924105	6.564782
		1.123081	0.019125	0.994003	1.059755	0.995494	6.065018
		–	–	0.260700	–	–	–
50.0	0.0786	1.433425	0.314432	0.416315	1.197257	0.574961	1.600836
		3.185976	0.111247	0.836114	1.784930	0.908598	6.095312
		1.183776	0.023924	0.992266	1.088015	0.994494	6.798272
		–	–	0.242610	–	–	–

APPENDIX C

N= 7   PRW= 2.0 DB (CONTINUED)

MSL	TW	A	B	C	WZ	WM	KM
55.0	0.1117	1.572778	0.312953	0.381555	1.254104	0.551713	1.627064
		3.689200	0.123250	0.810922	1.920729	0.893912	5.733045
		1.261895	0.028666	0.990513	1.123341	0.993533	7.519780
		-	-	0.227909	-	-	-
60.0	0.1521	1.741116	0.310528	0.352948	1.319514	0.531657	1.647673
		4.279443	0.133724	0.787920	2.068681	0.880327	5.448219
		1.359963	0.033213	0.988803	1.166175	0.992632	8.216633
		-	-	0.215876	-	-	-
65.0	0.2004	1.943044	0.307669	0.329401	1.393931	0.514439	1.663851
		4.972495	0.142765	0.767276	2.229909	0.867986	5.221030
		1.480903	0.037475	0.987178	1.216924	0.991803	8.878491
		-	-	0.205970	-	-	-
70.0	0.2569	2.184058	0.304694	0.309992	1.477856	0.499710	1.676562
		5.787016	0.150512	0.748984	2.405622	0.856926	5.037761
		1.628161	0.041397	0.985666	1.275994	0.991052	9.497788
		-	-	0.197778	-	-	-
75.0	0.3219	2.470708	0.301793	0.293961	1.571849	0.487142	1.686572
		6.745006	0.157117	0.732934	2.597115	0.847120	4.888593
		1.805833	0.044956	0.984282	1.343813	0.990381	10.069643
		-	-	0.190977	-	-	-
80.0	0.3961	2.810773	0.299072	0.280687	1.676536	0.476436	1.694483
		7.872388	0.162730	0.718952	2.805778	0.838498	4.766311
		2.018802	0.048147	0.983031	1.420846	0.989786	10.591573
		-	-	0.185312	-	-	-
85.0	0.4798	3.213478	0.296584	0.269670	1.792618	0.467329	1.700761
		9.199695	0.167488	0.706840	3.033100	0.830967	4.665485
		2.272884	0.050981	0.981913	1.507609	0.989262	11.063102
		-	-	0.180580	-	-	-
90.0	0.5736	3.689745	0.294349	0.260502	1.920871	0.459589	1.705766
		10.762895	0.171518	0.696394	3.280685	0.824423	4.581957
		2.574992	0.053480	0.980922	1.604678	0.988805	11.485330
		-	-	0.176619	-	-	-
95.0	0.6781	4.252438	0.292366	0.252856	2.062156	0.453014	1.709776
		12.604356	0.174927	0.687413	3.550261	0.818762	4.512490
		2.933331	0.055669	0.980050	1.712697	0.988407	11.860504
		-	-	0.173296	-	-	-
100.0	0.7941	4.916965	0.290623	0.246466	2.217423	0.447433	1.713004
		14.773999	0.177810	0.679713	3.843696	0.813879	4.454531
		3.357627	0.057576	0.979288	1.832383	0.988062	12.191651
		-	-	0.170505	-	-	-

N= 7   PRW= 3.0 DB

MSL	TW	A	B	C	WZ	WM	KM
35.0	0.0137	1.947336	0.243925	0.576581	1.395470	0.723165	2.287502
		1.126088	0.049122	0.924953	1.061173	0.955365	3.613734
		1.033848	0.006940	0.996478	1.016783	0.997582	5.290947
		-	-	0.263514	-	-	-

ELLIPTIC LOW-PASS FILTER DATA 213

N= 7   PRW= 3.0 DB (CONTINUED)

MSL	TW	A	B	C	WZ	WM	KM
40.0	0.0254	1.190991	0.252082	0.513589	1.091325	0.661226	1.772872
		2.238663	0.062408	0.896966	1.496216	0.944681	9.122131
		1.061054	0.010289	0.994598	1.030075	0.996475	6.147897
		–	–	0.238606	–	–	–
45.0	0.0425	1.274060	0.255789	0.460612	1.128743	0.627333	1.830382
		2.579873	0.074990	0.868304	1.606198	0.928789	8.270864
		1.100306	0.014007	0.992457	1.048955	0.995266	7.035749
		–	–	0.218823	–	–	–
50.0	0.0657	1.377775	0.256657	0.416609	1.173787	0.597627	1.876023
		2.979224	0.086497	0.840331	1.726043	0.913051	7.635680
		1.153728	0.017901	0.990174	1.074117	0.994020	7.934942
		–	–	0.202913	–	–	–
55.0	0.0955	1.505082	0.255807	0.380296	1.226818	0.571842	1.911984
		3.446897	0.096765	0.813967	1.856582	0.898000	7.148770
		1.223507	0.021803	0.987853	1.106123	0.992787	8.826644
		–	–	0.189998	–	–	–
60.0	0.1325	1.659515	0.253994	0.350411	1.288222	0.549616	1.940207
		3.995157	0.105768	0.789748	1.998789	0.883976	6.767868
		1.312029	0.025586	0.985577	1.145438	0.991603	9.694048
		–	–	0.179438	–	–	–
65.0	0.1771	1.845308	0.251707	0.325825	1.358421	0.530552	1.962328
		4.638605	0.113567	0.767917	2.153742	0.871163	6.465228
		1.422015	0.029161	0.983407	1.192483	0.990493	10.523241
		–	–	0.170752	–	–	–
70.0	0.2297	2.067529	0.249253	0.305577	1.437890	0.514260	1.979676
		5.394521	0.120266	0.748512	2.322611	0.859631	6.221841
		1.556654	0.032472	0.981381	1.247659	0.989473	11.303624
		–	–	0.163574	–	–	–
75.0	0.2908	2.332220	0.246818	0.288868	1.527161	0.500370	1.993310
		6.283299	0.125986	0.731445	2.506651	0.849372	6.024228
		1.719727	0.035492	0.979522	1.311384	0.988548	12.027943
		–	–	0.157618	–	–	–
80.0	0.3607	2.646568	0.244509	0.275048	1.626828	0.488549	2.004060
		7.328972	0.130851	0.716553	2.707207	0.840327	5.862552
		1.915736	0.038211	0.977841	1.384101	0.987720	12.692035
		–	–	0.152661	–	–	–
85.0	0.4399	3.019103	0.242382	0.263589	1.737557	0.478501	2.012571
		8.559849	0.134979	0.703637	2.925722	0.832411	5.729458
		2.150047	0.040635	0.976335	1.466304	0.986984	13.294393
		–	–	0.148522	–	–	–
90.0	0.5290	3.459928	0.240462	0.254063	1.860088	0.469968	2.019340
		10.009279	0.138475	0.692487	3.163745	0.825521	5.619344
		2.429044	0.042776	0.975000	1.558539	0.986337	13.835645
		–	–	0.145059	–	–	–
95.0	0.6285	3.980998	0.238752	0.246126	1.995244	0.462724	2.024750
		11.716544	0.141434	0.682895	3.422944	0.819552	5.527866
		2.760309	0.044656	0.973825	1.661418	0.985770	14.318031
		–	–	0.142155	–	–	–

APPENDIX C

N= 7   PRW= 3.0 DB (CONTINUED)

MSL	TW	A	B	C	hZ	WM	KM
100.0	0.7391	4.596441	0.237245	0.239498	2.143931	0.456577	2.029097
		13.727921	0.143936	0.674667	3.705121	0.814399	5.451611
		3.152835	0.046298	0.972796	1.775623	0.985276	14.744922
		–	–	0.139717	–	–	–

N= 8   PRW= 0.1 DB

MSL	TW	A	B	C	hZ	WM	KM
30.0	0.0227	1.052604	1.152436	0.576982	1.025965	–	–
		1.594174	0.382202	0.874711	1.262606	0.804470	1.358438
		8.276329	0.093121	0.985816	2.876861	0.990106	9.408968
		1.133635	0.016936	1.012252	1.064723	1.004848	6.430988
35.0	0.0364	1.083222	1.084425	0.488806	1.040780	–	–
		1.759678	0.422272	0.822693	1.326529	0.773513	1.370177
		9.760872	0.118048	0.974679	3.124239	0.982940	7.548186
		1.190093	0.023526	1.016212	1.090914	1.006333	6.328832
40.0	0.0545	1.124040	1.023866	0.421803	1.060208	–	–
		1.952097	0.453651	0.773201	1.397175	0.744594	1.375587
		11.446007	0.142774	0.962112	3.383195	0.974711	6.316040
		1.260222	0.030725	1.020368	1.122596	1.007913	6.322361
45.0	0.0776	1.176492	0.970961	0.370108	1.084662	–	–
		2.175178	0.477741	0.727665	1.474849	0.718009	1.376598
		13.364131	0.166549	0.948782	3.655698	0.965821	5.459761
		1.345698	0.038259	1.024578	1.160042	1.009549	6.374528
50.0	0.1059	1.242086	0.925185	0.329669	1.114489	–	–
		2.433375	0.495965	0.686706	1.559928	0.693873	1.374705
		15.553153	0.188873	0.935256	3.943749	0.956641	4.841790
		1.448443	0.045883	1.028728	1.203513	1.011203	6.462261
55.0	0.1399	1.322475	0.885770	0.297639	1.149989	–	–
		2.731921	0.509601	0.650425	1.652852	0.672175	1.371008
		18.057165	0.209456	0.921975	4.249372	0.947475	4.382179
		1.570682	0.053399	1.032730	1.253268	1.012842	6.570476
60.0	0.1797	1.419527	0.851911	0.271991	1.191439	–	–
		3.076931	0.519715	0.618618	1.754118	0.652816	1.366295
		20.927212	0.228170	0.909255	4.574627	0.938561	4.031929
		1.715015	0.060653	1.036621	1.309586	1.014437	6.689017
65.0	0.2256	1.535391	0.822852	0.251251	1.239109	–	–
		3.475506	0.527165	0.590927	1.864271	0.635647	1.361106
		24.222229	0.245002	0.897306	4.921608	0.930067	3.759653
		1.884477	0.067535	1.040063	1.372763	1.015966	6.810963
70.0	0.2780	1.672560	0.797916	0.234337	1.293275	–	–
		3.935879	0.532616	0.566929	1.983905	0.620492	1.355806
		28.010170	0.260016	0.886247	5.292464	0.922105	3.544488
		2.082612	0.073972	1.043332	1.443126	1.017411	6.931613
75.0	0.3371	1.833941	0.776512	0.220439	1.354231	–	–
		2.313548	0.536580	0.546192	1.521035	0.575151	1.243787
		32.369354	0.273321	0.876134	5.689407	0.914740	3.372114
		4.467579	0.079923	1.046321	2.113665	1.020381	9.820493
80.0	0.4032	2.022921	0.758131	0.208944	1.422294	–	–
		2.582086	0.539443	0.528305	1.606887	0.567916	1.252708
		37.390045	0.285054	0.866973	6.114740	0.908000	3.232434
		5.081621	0.085373	1.049031	2.254245	1.021517	9.539533

ELLIPTIC LOW-PASS FILTER DATA

N= 8   PRW= 0.1 DB (CONTINUED)

MSL	TW	A	B	C	WZ	WM	KM
85.0	0.4767	2.243447	0.742335	0.199379	1.497814	–	–
		2.893807	0.541496	0.512892	1.701119	0.561504	1.260292
		43.176280	0.295359	0.858737	6.570866	0.901884	3.118153
		5.790733	0.090323	1.051472	2.406394	1.022540	9.310168
90.0	0.5579	2.500112	0.728750	0.191381	1.581174	–	–
		3.255181	0.542955	0.499617	1.804212	0.555846	1.266744
		49.848047	0.304384	0.851380	7.060315	0.896376	3.023887
		6.609623	0.094787	1.053659	2.570919	1.023459	9.121442
95.0	0.6473	2.798249	0.717061	0.184662	1.672797	–	–
		3.673708	0.543980	0.488184	1.916692	0.550870	1.272238
		57.543668	0.312268	0.844840	7.585754	0.891445	2.945594
		7.555269	0.098791	1.055607	2.748685	1.024279	8.965100
100.0	0.7452	3.144061	0.706994	0.178996	1.773150	–	–
		4.158084	0.544689	0.478336	2.039138	0.546507	1.276921
		8.647314	0.319144	0.839051	2.940632	0.881806	2.650864
		66.422984	0.102364	1.057336	8.150030	1.025635	9.898371

N= 8   PRW= 0.5 DB

MSL	TW	A	B	C	WZ	WM	KM
30.0	0.0099	1.023607	0.833012	0.406055	1.011735	–	–
		6.442363	0.239808	0.835978	2.538181	0.893778	3.366539
		1.397614	0.048273	0.974432	1.182207	0.983829	6.249619
		1.073517	0.007500	1.001945	1.036107	1.000565	8.950811
35.0	0.0180	1.042056	0.782756	0.340472	1.020811	–	–
		7.683063	0.271670	0.782653	2.771834	0.858848	2.977380
		1.529450	0.064892	0.960695	1.236709	0.975454	5.675579
		1.112856	0.011397	1.002684	1.054920	1.000718	8.749956
40.0	0.0299	1.068639	0.737052	0.290632	1.033750	–	–
		9.087634	0.297048	0.731434	3.014570	0.824539	2.703640
		1.684067	0.082014	0.945126	1.297716	0.966030	5.259612
		1.163772	0.015925	1.003462	1.078783	1.000874	8.714994
45.0	0.0460	1.104804	0.696552	0.252223	1.051096	–	–
		1.864225	0.316752	0.684000	1.365366	0.761645	1.793690
		10.680790	0.098939	0.928536	3.268148	0.960579	8.909345
		1.227700	0.020883	1.004246	1.108016	1.001035	8.781177
50.0	0.0668	1.151973	0.661154	0.222237	1.073300	–	–
		2.073345	0.331767	0.641154	1.439911	0.735326	1.796430
		12.492319	0.115159	0.911641	3.534448	0.950774	7.704873
		1.306223	0.026077	1.005013	1.142901	1.001203	8.909136
55.0	0.0928	1.211611	0.630449	0.198551	1.100732	–	–
		2.315545	0.343045	0.603111	1.521692	0.711447	1.794578
		14.557487	0.130342	0.895012	3.815428	0.940962	6.832719
		1.401140	0.031331	1.005743	1.183698	1.001378	9.073471
60.0	0.1243	1.285297	0.603927	0.179641	1.133710	–	–
		2.595702	0.351421	0.569727	1.611118	0.690003	1.789834
		16.917537	0.144303	0.879065	4.113093	0.931407	6.182253
		1.514535	0.036505	1.006426	1.230665	1.001561	9.257156
65.0	0.1615	1.374798	0.581071	0.164401	1.172518	–	–
		2.919523	0.357581	0.540661	1.708663	0.670897	1.783409
		19.620343	0.156567	0.864076	4.429486	0.922296	5.685371
		1.648836	0.041491	1.007056	1.284070	1.001748	9.448551

APPENDIX C

N= 8   PRW= 0.5 DB (CONTINUED)

MSL	TW	A	B	C	WZ	WM	KM
70.0	0.2047	1.482129	0.561398	0.152014	1.217427	-	-
		3.293676	0.362013	0.515487	1.814349	0.653977	1.776145
		22.721246	0.168334	0.850208	4.766681	0.913755	5.298297
		1.906877	0.046213	1.007629	1.344201	1.001935	9.639674
75.0	0.2542	1.609626	0.544473	0.141870	1.268710	-	-
		3.725465	0.365320	0.493757	1.930250	0.639063	1.768611
		26.284092	0.173456	0.837532	5.126801	0.905854	4.991847
		1.991970	0.050622	1.008147	1.411372	1.002120	9.825125
80.0	0.3104	2.207977	0.529911	0.133506	1.485926	-	-
		4.225032	0.367645	0.475037	2.055488	0.625966	1.761177
		30.382475	0.187412	0.826059	5.512030	0.898625	4.745941
		1.760006	0.054691	1.008611	1.326652	1.001554	7.879112
85.0	0.3734	2.459391	0.517380	0.126568	1.568245	-	-
		4.801515	0.369292	0.458928	2.191236	0.614496	1.754073
		35.101210	0.195300	0.815756	5.924627	0.892070	4.546386
		1.936436	0.058410	1.009024	1.351559	1.001806	8.271357
90.0	0.4436	2.751440	0.506593	0.120783	1.658746	-	-
		5.467261	0.370444	0.445075	2.338218	0.604473	1.747434
		40.538056	0.202221	0.806561	6.366950	0.886169	4.382904
		2.142607	0.061783	1.009390	1.463765	1.002043	8.632133
95.0	0.5213	3.090186	0.497304	0.115936	1.757393	-	-
		6.236075	0.371235	0.433161	2.497213	0.595731	1.741327
		46.805716	0.208277	0.798397	6.841470	0.880889	4.247900
		2.382817	0.064820	1.009713	1.543638	1.002263	8.961491
100.0	0.6071	3.482663	0.489300	0.111859	1.866190	-	-
		7.123903	0.371767	0.422914	2.669064	0.588116	1.735777
		54.034176	0.213564	0.791178	7.350794	0.876188	4.135657
		2.662065	0.067540	1.009998	1.631584	1.002465	9.260185

N= 8   PRW= 1.0 DB

MSL	TW	A	B	C	WZ	WM	KM
35.0	0.0123	1.085935	0.647483	0.293721	1.042082	0.071063	1.000137
		6.855435	0.208930	0.780315	2.618289	0.367745	3.788486
		1.440902	0.045499	0.961047	1.200376	0.977690	7.223593
		1.029177	0.007423	1.000030	1.014484	0.999056	3.940643
40.0	0.0217	1.129122	0.609001	0.249212	1.062601	0.096392	1.000662
		8.150176	0.231332	0.727916	2.854851	0.834282	3.403410
		1.580334	0.059160	0.944858	1.257113	0.968465	6.655206
		1.050274	0.010823	0.999867	1.024829	0.998743	4.540179
45.0	0.0349	1.184375	0.574594	0.214969	1.088290	0.106537	1.001351
		9.616959	0.248881	0.679050	3.101122	0.802182	3.124500
		1.743446	0.072914	0.927355	1.320396	0.958480	6.229483
		1.080014	0.014659	0.999635	1.039237	0.998428	5.166781
50.0	0.0526	1.253170	0.544340	0.188288	1.119451	0.110619	1.002071
		11.282109	0.262335	0.634699	3.358885	0.772211	2.915615
		1.933214	0.086276	0.909339	1.390401	0.948179	5.899906
		1.119830	0.018764	0.999346	1.058220	0.998125	5.809465
55.0	0.0752	1.337156	0.517987	0.167257	1.156355	0.111760	1.002756
		2.153292	0.272486	0.595194	1.467410	0.728856	2.148853
		13.177228	0.098911	0.891466	3.630045	0.941205	8.915992
		1.171149	0.022989	0.999016	1.082196	0.997843	6.457194

ELLIPTIC LOW-PASS FILTER DATA

N= 8   PRW= 1.0 DB (CONTINUED)

MSL	TW	A	B	C	WZ	WM	KM
60.0	0.1031	1.438224	0.495155	0.150506	1.199260	0.111388	1.003378
		2.408051	0.280046	0.560457	1.551790	0.706232	2.144908
		15.339578	0.110616	0.874226	3.916577	0.931328	7.988484
		1.235467	0.027202	0.998661	1.111516	0.997585	7.099506
65.0	0.1365	1.558569	0.475437	0.137036	1.248427	0.110245	1.003926
		2.702647	0.285616	0.530179	1.643973	0.686004	2.138084
		17.812612	0.121294	0.857954	4.220499	0.921877	7.287591
		1.314422	0.031303	0.998293	1.146483	0.997354	7.727026
70.0	0.1758	1.700752	0.458438	0.126112	1.304129	0.108739	1.004400
		3.043105	0.289681	0.503941	1.744450	0.668043	2.129666
		20.646684	0.130920	0.842849	4.543862	0.912992	6.746390
		1.409858	0.035217	0.997926	1.187374	0.997149	8.331808
75.0	0.2211	1.867765	0.443795	0.117185	1.366662	0.107101	1.004806
		3.436432	0.292620	0.481290	1.853762	0.652182	2.120537
		23.899956	0.139518	0.829013	4.888758	0.904757	6.321017
		1.523895	0.038895	0.997567	1.234461	0.996970	8.907524
80.0	0.2729	2.063103	0.431187	0.109840	1.436350	0.105460	1.005151
		3.890750	0.294723	0.461781	1.972498	0.638233	2.111285
		27.639497	0.147145	0.816469	5.257328	0.897212	5.981731
		1.658992	0.042307	0.997224	1.288018	0.996813	9.449501
85.0	0.3314	2.290834	0.420332	0.103759	1.513550	0.103889	1.005443
		4.415460	0.296211	0.445000	2.101300	0.626005	2.102286
		31.942611	0.153874	0.805191	5.651779	0.890362	5.707772
		1.818014	0.045439	0.996901	1.348337	0.996678	9.954641
90.0	0.3969	2.555697	0.410983	0.098698	1.598655	0.102427	1.005689
		5.021431	0.297249	0.430575	2.240855	0.615312	2.093771
		36.898389	0.159787	0.795116	6.074404	0.884190	5.484277
		2.004302	0.048287	0.996601	1.415734	0.996560	10.421259
95.0	0.4696	2.863193	0.402929	0.094464	1.692097	0.101090	1.005897
		5.721225	0.297961	0.418176	2.391908	0.605980	2.085867
		42.609527	0.164966	0.786166	6.527597	0.878665	5.300364
		2.221748	0.050860	0.996325	1.490553	0.996460	10.848880
100.0	0.5501	3.219709	0.395988	0.090908	1.794355	0.099883	1.006072
		6.529356	0.298436	0.407518	2.555261	0.597847	2.078633
		49.194459	0.169491	0.778247	7.013876	0.873743	5.147912
		2.474881	0.053169	0.996073	1.573176	0.996373	11.238010

N= 8   PRW= 2.0 DB

MSL	TW	A	B	C	WZ	WM	KM
45.0	0.0251	1.143957	0.449520	0.187534	1.069559	0.245596	1.054649
		8.559948	0.182371	0.684845	2.925739	0.815698	4.209898
		1.625431	0.049226	0.930969	1.274924	0.962557	8.411270
		1.058006	0.009304	0.997082	1.028594	0.997809	6.255836
50.0	0.0396	1.203008	0.424920	0.163050	1.096817	0.232240	1.058588
		10.081710	0.194040	0.638390	3.175171	0.785533	3.893955
		1.795992	0.059567	0.912168	1.340146	0.952230	7.926314
		1.090544	0.012312	0.996019	1.044291	0.997169	7.090557
55.0	0.0587	1.276061	0.403374	0.143816	1.129629	0.220529	1.061828
		11.810533	0.202904	0.596771	3.436646	0.757652	3.653517
		1.994206	0.069484	0.893255	1.412164	0.941774	7.544149
		1.133571	0.015479	0.994875	1.064693	0.996513	7.940062

APPENDIX C

N= 8   PRW= 2.0 DB (CONTINUED)

MSL	TW	A	B	C	WZ	WM	KM
60.0	0.0823	1.364830	0.384636	0.128547	1.168260	0.210323	1.064489
		13.779615	0.209540	0.560034	3.712090	0.732298	3.466326
		2.223931	0.078771	0.874821	1.491285	0.931510	7.236765
		1.188528	0.018692	0.993694	1.090197	0.995861	8.790259
65.0	0.1123	1.471289	0.368410	0.116308	1.212967	0.201463	1.066674
		16.027958	0.214449	0.527934	4.003493	0.709514	3.317920
		2.489760	0.087310	0.857284	1.577897	0.921676	6.985711
		1.256947	0.021863	0.992513	1.121136	0.995229	9.628043
70.0	0.1474	1.597731	0.354396	0.106414	1.264014	0.193783	1.068469
		2.797093	0.218042	0.500076	1.672451	0.682865	2.729375
		18.600950	0.095054	0.840909	4.312882	0.914311	9.225952
		1.340515	0.024923	0.991360	1.157806	0.994629	10.441893
75.0	0.1884	1.746839	0.342309	0.098354	1.321631	0.187151	1.069944
		3.152223	0.220646	0.476009	1.775451	0.665812	2.717966
		21.551122	0.102003	0.825842	4.642319	0.905664	8.583459
		1.441153	0.027822	0.990257	1.200480	0.994063	11.222259
80.0	0.2356	1.921744	0.331391	0.091740	1.386270	0.181417	1.071159
		3.562482	0.222513	0.455276	1.887454	0.650783	2.705927
		24.939115	0.108189	0.812136	4.993908	0.897714	8.075053
		1.551072	0.030530	0.989219	1.249429	0.993550	11.961731
85.0	0.2893	2.126098	0.322916	0.086278	1.458114	0.176465	1.072161
		4.036355	0.223836	0.437443	2.009063	0.637587	2.693919
		28.834843	0.113661	0.799780	5.369808	0.890479	7.667245
		1.702344	0.033029	0.988255	1.304931	0.993077	12.655029
90.0	0.3498	2.364157	0.315183	0.081744	1.537581	0.172189	1.072989
		4.583594	0.224761	0.422118	2.140933	0.626034	2.682361
		33.318831	0.118478	0.788721	5.772251	0.883948	7.336392
		1.869465	0.035314	0.987369	1.367284	0.992648	13.298859
95.0	0.4173	2.640871	0.308520	0.077959	1.625076	0.168498	1.073676
		5.215599	0.225396	0.408952	2.283769	0.615942	2.671501
		33.484059	0.122704	0.778881	6.203555	0.878091	7.065399
		2.064428	0.037384	0.986562	1.436812	0.992263	13.891676
100.0	0.4923	2.961933	0.302776	0.074786	1.721043	0.165311	1.074247
		5.945454	0.225821	0.397640	2.438330	0.607141	2.661472
		44.437549	0.126401	0.770166	6.666150	0.872867	6.841641
		2.291803	0.039248	0.985832	1.513870	0.991917	14.433399

N= 8   PRW= 3.0 DB

MSL	TW	A	B	C	WZ	WM	KM
50.0	0.0324	1.174099	0.353887	0.153179	1.083558	0.278542	1.151677
		9.355162	0.154959	0.646685	3.058621	0.795559	4.860572
		1.714012	0.045130	0.916545	1.309203	0.955606	9.888855
		1.074303	0.008993	0.995063	1.036486	0.996999	8.238382
55.0	0.0493	1.240487	0.335389	0.134414	1.113771	0.260547	1.156885
		10.984584	0.162961	0.603407	3.314300	0.767206	4.536019
		1.899021	0.053428	0.897173	1.378050	0.945090	9.381785
		1.112310	0.011534	0.993573	1.054661	0.996190	9.273803
60.0	0.0710	1.321772	0.319255	0.119558	1.149683	0.246870	1.161119
		12.838247	0.168979	0.565062	3.583050	0.741275	4.285052
		2.113672	0.061218	0.878113	1.453847	0.934657	8.975891
		1.161576	0.014149	0.992021	1.077764	0.995365	10.316885

ELLIPTIC LOW-PASS FILTER DATA

N= 8   PRW= 3.0 DB (CONTINUED)

MSL	TW	A	B	C	WZ	WM	KM
65.0	0.0980	1.419798	0.305257	0.107683	1.191553	0.235170	1.164567
		14.952411	0.173447	0.531472	3.866835	0.717871	4.087152
		2.362210	0.068428	0.859851	1.536948	0.924574	8.645530
		1.223581	0.016758	0.990460	1.106156	0.994549	11.351066
70.0	0.1304	1.536705	0.293151	0.098108	1.239639	0.225154	1.167380
		17.369429	0.176727	0.502276	4.167665	0.696954	3.928628
		2.649654	0.074998	0.842706	1.627776	0.915029	8.373253
		1.299931	0.019296	0.988929	1.140145	0.993761	12.361425
75.0	0.1687	1.674988	0.282700	0.090324	1.294213	0.216574	1.169679
		20.138429	0.179110	0.477028	4.487586	0.678397	3.800017
		2.981873	0.080916	0.826864	1.726810	0.906138	8.146694
		1.392432	0.021717	0.987460	1.180014	0.993013	13.335272
80.0	0.2130	1.837561	0.273687	0.083952	1.355567	0.209217	1.171564
		23.316173	0.180823	0.455266	4.828682	0.662026	3.694581
		3.365698	0.086199	0.812407	1.834584	0.897962	7.958781
		1.503152	0.023991	0.986073	1.226031	0.992315	14.262463
85.0	0.2637	2.027827	0.265918	0.078701	1.424018	0.202904	1.173111
		3.809052	0.182041	0.436546	1.951679	0.644817	3.262413
		26.968137	0.090882	0.799341	5.193085	0.891605	9.559863
		1.634489	0.026099	0.984782	1.278471	0.991670	15.135482
90.0	0.3210	2.249753	0.259224	0.074348	1.499918	0.197483	1.174386
		4.321106	0.182895	0.420459	2.078727	0.632626	3.248019
		31.169785	0.095012	0.787625	5.582991	0.884804	9.119154
		1.789234	0.028032	0.983593	1.337623	0.991081	15.949326
95.0	0.3852	2.507956	0.253454	0.070722	1.583653	0.192823	1.175440
		4.912466	0.183483	0.406641	2.216408	0.621968	3.234363
		36.008088	0.098639	0.777184	6.000674	0.878695	8.759464
		1.970643	0.029790	0.982509	1.403796	0.990547	16.701258
100.0	0.4568	2.807801	0.248430	0.067687	1.675650	0.188815	1.176313
		5.595391	0.183879	0.394772	2.365458	0.612669	3.221664
		41.583300	0.101815	0.767926	6.448511	0.873240	8.463360
		2.182506	0.031376	0.981528	1.477331	0.990066	17.390482

N= 9   PRW= 0.1 DB

MSL	TW	A	B	C	WZ	WM	KM
30.0	0.0108	1.024902	0.710230	0.737736	1.012374	–	–
		2.581611	0.191148	0.937246	1.606739	0.947935	3.297903
		1.249924	0.044861	0.993162	1.118000	0.992172	4.650831
		1.062018	0.008100	1.005860	1.030543	1.002324	6.619376
		–	–	0.691918	–	–	–
35.0	0.0183	1.041481	0.720954	0.657480	1.020530	–	–
		2.933715	0.226876	0.904529	1.712809	0.925450	2.973438
		1.331320	0.060428	0.987026	1.153828	0.987318	4.340104
		1.092493	0.011914	1.008210	1.045224	1.003216	6.572436
		–	–	0.625952	–	–	–
40.0	0.0288	1.064455	0.721097	0.588635	1.031724	0.067336	1.000021
		1.426887	0.259110	0.870059	1.194524	0.859918	1.622991
		3.329161	0.077005	0.979537	1.824599	0.986968	9.094092
		1.131465	0.016337	1.010828	1.063703	1.004222	6.629442
		–	–	0.573485	–	–	–

APPENDIX C

N= 9   PRW= 0.1 DB (CONTINUED)

MSL	TW	A	B	C	WZ	WM	KM
45.0	0.0425	1.094845	0.714756	0.530307	1.046348	0.155054	1.000788
		1.538049	0.287476	0.835430	1.240181	0.838807	1.649263
		3.773479	0.093994	0.971038	1.942544	0.981616	7.811346
		1.179925	0.021216	1.013626	1.086244	1.005316	6.752656
		-	-	0.530938	-	-	-
50.0	0.0600	1.133648	0.704710	0.481182	1.064729	0.189833	1.002309
		1.666513	0.311990	0.801814	1.290935	0.818712	1.668326
		4.273226	0.110896	0.961885	2.067178	0.975798	6.880610
		1.238924	0.026392	1.016517	1.113069	1.006475	6.917995
		-	-	0.495929	-	-	-
55.0	0.0814	1.181877	0.692793	0.439897	1.087142	0.208693	1.004255
		1.814298	0.332902	0.769993	1.346959	0.799861	1.681692
		4.836062	0.127323	0.952404	2.199105	0.969716	6.184523
		1.309609	0.031719	1.019429	1.144382	1.007670	7.109135
		-	-	0.466799	-	-	-
60.0	0.1070	1.240602	0.680181	0.405193	1.113823	0.219771	1.006417
		1.983765	0.350582	0.740428	1.408462	0.782382	1.690660
		5.470828	0.142994	0.942876	2.338980	0.963549	5.650926
		1.393269	0.037066	1.022300	1.180368	1.008878	7.314540
		-	-	0.442346	-	-	-
65.0	0.1371	1.310986	0.667609	0.375974	1.144983	0.226471	1.008664
		2.177646	0.365441	0.713338	1.475685	0.766327	1.696302
		6.187658	0.157727	0.933521	2.487500	0.957446	5.233490
		1.491364	0.042329	1.025081	1.221214	1.010077	7.525865
		-	-	0.421674	-	-	-
70.0	0.1718	1.394325	0.655521	0.351314	1.180816	0.230531	1.010907
		2.399086	0.377884	0.688774	1.548898	0.751688	1.699478
		6.998109	0.171413	0.924507	2.645394	0.951520	4.901362
		1.605564	0.047423	1.027739	1.267108	1.011249	7.737013
		-	-	0.404099	-	-	-
75.0	0.2113	1.492089	0.644165	0.330442	1.221511	0.232936	1.013089
		2.651694	0.388284	0.666673	1.628402	0.738422	1.700865
		7.915329	0.184006	0.915949	2.813420	0.945854	4.633341
		1.737784	0.052288	1.030249	1.318250	1.012379	7.943554
		-	-	0.389085	-	-	-
80.0	0.2558	1.605949	0.633665	0.312723	1.267261	0.234281	1.015175
		2.939593	0.396975	0.646905	1.714524	0.726458	1.700983
		8.954259	0.195505	0.907923	2.992367	0.940507	4.414456
		1.890221	0.056881	1.032595	1.374853	1.013456	8.142337
		-	-	0.376207	-	-	-
85.0	0.3056	1.737822	0.624067	0.297634	1.318265	0.234941	1.017140
		3.267490	0.404240	0.629301	1.807620	0.715711	1.700228
		10.131858	0.205939	0.900469	3.183058	0.935513	4.233883
		2.065393	0.061176	1.034771	1.437148	1.014472	8.331195
		-	-	0.365126	-	-	-
90.0	0.3607	1.889902	0.615365	0.284748	1.374737	0.235159	1.018972
		3.640751	0.410322	0.613677	1.908075	0.706090	1.698898
		11.467384	0.215357	0.893605	3.386353	0.930889	4.083625
		2.266186	0.065160	1.036774	1.505386	1.015421	8.508741
		-	-	0.355563	-	-	-
95.0	0.4215	2.064706	0.607524	0.273710	1.436908	0.235094	1.020667
		4.065487	0.415423	0.599843	2.016305	0.697500	1.697213
		12.982682	0.223824	0.887324	3.603149	0.926639	3.957669
		2.495898	0.068829	1.038607	1.579841	1.016303	8.674190
		-	-	0.347290	-	-	-

ELLIPTIC LOW-PASS FILTER DATA 221

N= 9   PRW= 0.1 DB (CONTINUED)

MSL	TW	A	B	C	WZ	WM	KM
100.0	0.4882	2.265117	0.600492	0.264231	1.505031	0.234854	1.022222
		4.548665	0.419712	0.587617	2.132760	0.689847	1.695331
		14.702604	0.231409	0.881611	3.834397	0.922756	3.851413
		2.758307	0.072188	1.040277	1.660815	1.017116	8.827227
		—	—	0.340117	—	—	—

N= 9   PRW= 0.5 DB

MSL	TW	A	B	C	WZ	WM	KM
40.0	0.0147	1.033493	0.501673	0.515271	1.016609	0.461161	1.074378
		2.774438	0.159478	0.855674	1.665665	0.912016	4.064835
		1.293996	0.041282	0.972372	1.137539	0.983048	6.001258
		1.078156	0.007896	1.001716	1.038343	1.000434	9.038601
		—	—	0.420753	—	—	—
45.0	0.0238	1.053532	0.499263	0.460937	1.026417	0.444712	1.087400
		3.150176	0.181100	0.819048	1.774874	0.889562	3.751818
		1.383172	0.052683	0.961930	1.176083	0.976847	5.732707
		1.113275	0.010916	1.002219	1.055119	1.000542	9.198296
		—	—	0.387530	—	—	—
50.0	0.0361	1.080546	0.493270	0.415105	1.039493	0.428871	1.099200
		3.572294	0.200039	0.783126	1.890051	0.867240	3.508123
		1.487298	0.064365	0.950579	1.219548	0.970158	5.527442
		1.157448	0.014262	1.002740	1.075848	1.000660	9.431223
		—	—	0.360128	—	—	—
55.0	0.0519	1.115537	0.485285	0.376572	1.056190	0.414114	1.109880
		1.607944	0.216345	0.748866	1.268047	0.828472	2.261117
		4.046855	0.075979	0.938736	2.011680	0.966494	9.813394
		1.211691	0.017822	1.003265	1.100768	1.000789	9.710394
		—	—	0.337281	—	—	—
60.0	0.0715	1.277103	0.476364	0.344197	1.130090	0.410358	1.136579
		1.746988	0.230215	0.716863	1.321737	0.809365	2.282449
		4.581012	0.087254	0.926767	2.140330	0.959706	8.821659
		1.159507	0.021492	1.003779	1.076804	1.000288	6.330045
		—	—	0.318072	—	—	—
65.0	0.0953	1.354906	0.467187	0.316970	1.164004	0.397376	1.145445
		1.906636	0.241914	0.687432	1.380810	0.791601	2.296947
		5.183074	0.097998	0.914969	2.276637	0.952951	8.059265
		1.213493	0.025178	1.004275	1.101587	1.000460	6.922795
		—	—	0.301813	—	—	—
70.0	0.1234	1.446480	0.458182	0.294029	1.202697	0.385664	1.153117
		2.089451	0.251729	0.660680	1.445493	0.775253	2.306268
		5.862592	0.108087	0.903569	2.421279	0.946366	7.461425
		1.278615	0.028806	1.004746	1.130758	1.000646	7.506583
		—	—	0.287976	—	—	—
75.0	0.1560	1.553399	0.449602	0.274649	1.246354	0.375161	1.159744
		2.298394	0.259935	0.636577	1.516046	0.760329	2.311738
		6.630478	0.117448	0.892726	2.574971	0.940050	6.984811
		1.356105	0.032317	1.005187	1.164519	1.000839	8.075105
		—	—	0.276147	—	—	—
80.0	0.1934	1.677466	0.441588	0.258232	1.295170	0.365787	1.165460
		2.536862	0.266786	0.615002	1.592753	0.746793	2.314402
		7.499162	0.126051	0.882545	2.738460	0.934074	6.599540
		1.447352	0.035667	1.005596	1.203059	1.001034	8.623062
		—	—	0.265997	—	—	—

APPENDIX C

N= 9   PRW= 0.5 DB (CONTINUED)

MSL	TW	A	B	C	WZ	WM	KM
85.0	0.2357	1.820750	0.434206	0.244283	1.349352	0.357447	1.170387
		2.808746	0.272505	0.595786	1.675931	0.734579	2.315064
		8.482771	0.133899	0.873085	2.912520	0.928481	6.284445
		1.553935	0.038827	1.005973	1.246569	1.001228	9.146218
		–	–	0.257261	–	–	–
90.0	0.2831	1.985623	0.427474	0.232396	1.409121	0.350046	1.174635
		3.118488	0.277281	0.578733	1.765924	0.723606	2.314336
		9.597350	0.141013	0.864370	3.097959	0.923295	6.024175
		1.677658	0.041780	1.006318	1.295244	1.001415	9.641404
		–	–	0.249719	–	–	–
95.0	0.3359	2.174803	0.421381	0.222237	1.474721	0.343492	1.178298
		3.471155	0.281276	0.563640	1.863104	0.713780	2.312680
		10.861110	0.147429	0.856398	3.295620	0.918522	5.807368
		1.820589	0.044516	1.006631	1.349292	1.001595	10.106458
		–	–	0.243194	–	–	–
100.0	0.3941	2.391403	0.415897	0.213531	1.546416	0.337694	1.181458
		3.872522	0.284624	0.550309	1.967872	0.705007	2.310435
		12.294729	0.153192	0.849147	3.506384	0.914158	5.625455
		1.985099	0.047033	1.006914	1.408936	1.001766	10.540130
		–	–	0.237538	–	–	–

N= 9   PRW= 1.0 DB

MSL	TW	A	B	C	WZ	WM	KM
45.0	0.0175	1.089362	0.404447	0.445245	1.043725	0.521861	1.223893
		2.899727	0.137978	0.821891	1.702858	0.897173	4.751958
		1.323292	0.037573	0.962557	1.150344	0.978822	7.170500
		1.039711	0.007417	0.999815	1.019662	0.999205	5.265533
		–	–	0.322781	–	–	–
50.0	0.0277	1.127506	0.400382	0.399503	1.061841	0.498051	1.244443
		3.290908	0.154179	0.784965	1.814086	0.875057	4.419704
		1.417487	0.046900	0.950696	1.190582	0.972178	6.897040
		1.062052	0.009982	0.999652	1.030559	0.999003	5.964237
		–	–	0.299077	–	–	–
55.0	0.0411	1.175050	0.394304	0.361027	1.083997	0.476573	1.262205
		3.730454	0.168222	0.749498	1.931438	0.853425	4.156439
		1.527143	0.056307	0.938174	1.235776	0.965172	6.683260
		1.091717	0.012770	0.999453	1.044853	0.998902	6.685207
		–	–	0.279316	–	–	–
60.0	0.0582	1.233033	0.387196	0.328703	1.110420	0.457339	1.277511
		4.224797	0.180220	0.716196	2.055431	0.832725	3.943874
		1.653935	0.065541	0.925405	1.286054	0.958031	6.512090
		1.129705	0.015690	0.999226	1.062876	0.998609	7.418578
		–	–	0.262702	–	–	–
65.0	0.0793	1.302593	0.379705	0.301534	1.141312	0.440202	1.290673
		4.781481	0.190372	0.685452	2.186660	0.813248	3.769701
		1.799851	0.074417	0.912732	1.341585	0.950942	6.372437
		1.177024	0.018661	0.998980	1.084907	0.998426	8.154595
		–	–	0.248641	–	–	–
70.0	0.1045	1.385002	0.372243	0.278659	1.176861	0.424989	1.301975
		1.967217	0.198903	0.657430	1.402575	0.786372	2.796128
		5.409234	0.082808	0.900420	2.325776	0.946375	9.569447
		1.234737	0.021614	0.998722	1.111187	0.998255	8.884043
		–	–	0.236676	–	–	–

ELLIPTIC LOW-PASS FILTER DATA

N= 9   PRW= 1.0 DB (CONTINUED)

MSL	TW	A	B	C	WZ	WM	KM
75.0	0.1342	1.481704	0.365063	0.259356	1.217253	0.411521	1.311673
		2.158729	0.206043	0.632134	1.469261	0.770623	2.805593
		6.118063	0.090635	0.888662	2.473472	0.939767	8.908642
		1.303998	0.024496	0.998461	1.141927	0.998099	9.598591
		−	−	0.226450	−	−	−
80.0	0.1684	1.594348	0.358308	0.243022	1.262675	0.399623	1.319991
		2.377494	0.212005	0.609461	1.541912	0.756288	2.810853
		6.919386	0.097859	0.877586	2.630473	0.933494	8.377771
		1.386088	0.027264	0.998201	1.177323	0.997957	10.291037
		−	−	0.217676	−	−	−
85.0	0.2075	1.724825	0.352055	0.229160	1.313326	0.389127	1.327129
		2.627074	0.216979	0.589249	1.620825	0.743317	2.813074
		7.826200	0.104471	0.867270	2.797535	0.927609	7.945856
		1.482460	0.029889	0.997948	1.217563	0.997829	10.955453
		−	−	0.210125	−	−	−
90.0	0.2515	1.875304	0.346331	0.217362	1.369417	0.379878	1.333256
		2.911542	0.221130	0.571302	1.706324	0.731637	2.813151
		8.853269	0.110480	0.857747	2.975444	0.922139	7.590668
		1.594765	0.032353	0.997704	1.262840	0.997714	11.587227
		−	−	0.203609	−	−	−
95.0	0.3008	2.048273	0.341134	0.207292	1.431179	0.371734	1.338520
		3.235550	0.224598	0.555414	1.798763	0.721160	2.811759
		10.017358	0.115911	0.849023	3.165021	0.917097	7.295908
		1.724895	0.034644	0.997474	1.313353	0.997612	12.183031
		−	−	0.197972	−	−	−
100.0	0.3554	2.246581	0.336447	0.198671	1.498860	0.364567	1.343048
		3.604401	0.227500	0.541379	1.898526	0.711791	2.809402
		11.337504	0.120798	0.841079	3.367121	0.912480	7.049385
		1.875019	0.036759	0.997257	1.369313	0.997521	12.740731
		−	−	0.193086	−	−	−

N= 9   PRW= 2.0 DB

MSL	TW	A	B	C	WZ	WM	KM
55.0	0.0312	1.140382	0.302228	0.354719	1.067887	0.522094	1.517414
		3.414076	0.121826	0.757501	1.847722	0.863644	5.592672
		1.447860	0.038405	0.940976	1.203271	0.968249	8.881013
		1.069918	0.008353	0.997234	1.034368	0.998119	8.179114
		−	−	0.219636	−	−	−
60.0	0.0457	1.190864	0.297056	0.321509	1.091267	0.498458	1.540984
		3.868919	0.131745	0.722712	1.966957	0.842664	5.281077
		1.562337	0.045507	0.927636	1.249935	0.961031	8.634907
		1.101912	0.010515	0.996452	1.049720	0.997673	9.135074
		−	−	0.205878	−	−	−
65.0	0.0639	1.252107	0.291384	0.293608	1.118976	0.477547	1.561132
		4.380646	0.140176	0.690395	2.092999	0.822758	5.027048
		1.694497	0.052414	0.914246	1.301728	0.953762	8.435265
		1.142520	0.012754	0.995631	1.068887	0.997220	10.102079
		−	−	0.194244	−	−	−
70.0	0.0862	1.325283	0.285601	0.270139	1.151209	0.459093	1.578348
		4.957139	0.147282	0.660802	2.226463	0.804149	4.817303
		1.846424	0.059002	0.901122	1.358832	0.946611	8.270443
		1.192764	0.015009	0.994794	1.092137	0.996770	11.067631
		−	−	0.184351	−	−	−

224 APPENDIX C

N= 9   PRW= 2.0 DB (CONTINUED)

MSL	TW	A	B	C	WZ	WM	KM
75.0	0.1127	1.411711	0.279952	0.250357	1.188154	0.442839	1.593056
		5.607487	0.153240	0.633996	2.368013	0.786954	4.642355
		2.020549	0.065193	0.888502	1.421460	0.939708	8.132691
		1.253727	0.017233	0.993962	1.119699	0.996332	12.020097
		−	−	0.175902	−	−	−
80.0	0.1437	1.512890	0.274584	0.233642	1.229996	0.428540	1.605627
		6.342094	0.158218	0.609912	2.518351	0.771210	4.495230
		2.219684	0.070940	0.876549	1.489860	0.933145	8.016420
		1.326596	0.019389	0.993148	1.151780	0.995912	12.949121
		−	−	0.168658	−	−	−
85.0	0.1793	1.630538	0.269577	0.219478	1.276925	0.415974	1.616376
		7.172823	0.162371	0.588405	2.678213	0.756900	4.370670
		2.447063	0.076223	0.865368	1.564309	0.926982	7.917521
		1.412702	0.021448	0.992366	1.188571	0.995515	13.845901
		−	−	0.162427	−	−	−
90.0	0.2198	1.766622	0.264969	0.207440	1.329143	0.404936	1.625576
		8.113164	0.165835	0.569285	2.848362	0.743968	4.264627
		2.706391	0.081042	0.855011	1.645111	0.921252	7.832886
		1.513548	0.023392	0.991623	1.230263	0.995144	14.703326
		−	−	0.157052	−	−	−
95.0	0.2654	1.923400	0.260769	0.197180	1.386867	0.395246	1.633459
		3.001904	0.168726	0.552346	1.732600	0.729254	3.643347
		9.178441	0.085410	0.845497	3.029594	0.917121	9.789322
		1.630854	0.025208	0.990926	1.277049	0.994799	15.516003
		−	−	0.152406	−	−	−
100.0	0.3162	2.103460	0.256968	0.188410	1.450331	0.386739	1.640223
		3.338435	0.171143	0.537375	1.827139	0.719177	3.641059
		10.386053	0.089350	0.836815	3.222740	0.912209	9.428709
		1.766589	0.026892	0.990277	1.329131	0.994482	16.280180
		−	−	0.148380	−	−	−

N= 9   PRW= 3.0 DB

MSL	TW	A	B	C	WZ	WM	KM
60.0	0.0386	1.507397	0.244072	0.322307	1.227761	0.526732	1.931396
		3.652274	0.104377	0.730817	1.911092	0.850094	6.578942
		1.166285	0.034718	0.930842	1.079947	0.962023	5.681625
		1.036130	0.007818	0.995591	1.042175	0.997441	10.682238
		−	−	0.171978	−	−	−
65.0	0.0550	1.631153	0.239518	0.293425	1.277166	0.503014	1.950897
		4.136826	0.111702	0.697419	2.033919	0.829859	6.243421
		1.222421	0.040428	0.917137	1.105632	0.954686	5.980218
		1.122636	0.009624	0.994520	1.059545	0.996873	11.862999
		−	−	0.161884	−	−	−
70.0	0.0754	1.773673	0.234775	0.269144	1.331793	0.482076	1.965448
		4.682361	0.117892	0.666708	2.163876	0.810844	5.967369
		1.289933	0.045915	0.903601	1.135752	0.947435	6.258499
		1.168301	0.011463	0.993422	1.080880	0.996298	13.048197
		−	−	0.153308	−	−	−
75.0	0.0999	1.937223	0.230081	0.248695	1.391842	0.463636	1.976188
		5.297398	0.123091	0.638805	2.301608	0.793200	5.737767
		1.370067	0.051100	0.890507	1.170499	0.940409	6.516389
		1.224174	0.013292	0.992323	1.106424	0.995730	14.223120
		−	−	0.145990	−	−	−

ELLIPTIC LOW-PASS FILTER DATA

N= 9   PRW= 3.0 DB (CONTINUED)

MSL	TW	A	B	C	WZ	WM	KM
80.0	0.1288	1.464236	0.225582	0.231432	1.210056	0.444008	1.891736
		5.991719	0.127439	0.613677	2.447799	0.776993	5.545123
		2.124435	0.055935	0.878047	1.457544	0.935031	9.853822
		1.291390	0.015076	0.991245	1.136393	0.995179	15.374411
		-	-	0.139720	-	-	-
85.0	0.1623	1.574052	0.221360	0.216818	1.254612	0.430190	1.905991
		6.776493	0.131068	0.591200	2.603170	0.762222	5.382331
		2.338343	0.060396	0.866345	1.529164	0.928644	9.725669
		1.371209	0.016790	0.990205	1.170986	0.994651	16.490485
		-	-	0.134330	-	-	-
90.0	0.2005	1.701361	0.217456	0.204413	1.304362	0.418067	1.918174
		7.664435	0.134093	0.571196	2.768472	0.748846	5.243956
		2.582426	0.064476	0.855475	1.606993	0.922682	9.616100
		1.465049	0.018415	0.989215	1.210392	0.994153	17.561766
		-	-	0.129684	-	-	-
95.0	0.2437	1.848283	0.213887	0.193850	1.359516	0.407433	1.928598
		8.669996	0.136618	0.553457	2.944486	0.736793	5.125759
		2.860669	0.068184	0.845464	1.691351	0.917166	9.521941
		1.574527	0.019940	0.988283	1.254801	0.993687	18.580778
		-	-	0.125669	-	-	-
100.0	0.2921	2.017250	0.210648	0.184832	1.420299	0.398106	1.937530
		9.809584	0.138726	0.537771	3.132025	0.725977	5.024383
		3.177620	0.071534	0.836311	1.782588	0.912102	9.440699
		1.701491	0.021357	0.987415	1.304412	0.993255	19.542101
		-	-	0.122192	-	-	-

N= 10   PRW= 0.1 DB

MSL	TW	A	B	C	WZ	WM	KM
40.0	0.0153	1.033932	1.004805	0.406185	1.016824	-	-
		1.877404	0.445994	0.744840	1.370184	0.730177	1.373083
		11.015717	0.143281	0.928083	3.318993	0.957049	6.181142
		1.208809	0.041343	0.989009	1.099458	0.990200	4.467407
		1.068201	0.008706	1.005770	1.033538	1.002253	6.802416
45.0	0.0235	1.051817	0.946599	0.351684	1.025581	-	-
		1.270690	0.467002	0.691892	1.127249	0.617057	1.142635
		12.684060	0.167112	0.904209	3.561469	0.942402	5.311830
		2.062153	0.052757	0.983734	1.436020	0.989853	9.859423
		1.096617	0.011793	1.007573	1.047195	1.002961	6.977590
50.0	0.0343	1.075290	0.895249	0.308566	1.036962	-	-
		1.342895	0.481755	0.643434	1.158834	0.600654	1.163317
		14.526923	0.189440	0.879434	3.811420	0.926933	4.680956
		2.268925	0.064713	0.977738	1.506295	0.986145	8.724765
		1.131964	0.015220	1.009524	1.063938	1.003743	7.202867
55.0	0.0479	1.105084	0.850121	0.274016	1.051230	-	-
		1.426364	0.491644	0.599776	1.194305	0.584226	1.180794
		16.565465	0.209983	0.854549	4.070069	0.911119	4.208651
		2.500116	0.076892	0.971220	1.581175	0.982100	7.868539
		1.174969	0.018893	1.011570	1.083960	1.004582	7.461454
60.0	0.0646	1.141925	0.810523	0.246022	1.068609	-	-
		1.522187	0.497847	0.560858	1.233769	0.568326	1.195350
		18.824215	0.228608	0.830167	4.338688	0.895358	3.845976
		2.758532	0.089021	0.964377	1.660883	0.977841	7.206566
		1.226400	0.022719	1.013664	1.107429	1.005460	7.741082

APPENDIX C

N= 10   PRW= 0.1 DB (CONTINUED)

MSL	TW	A	B	C	WZ	WM	KM
65.0	0.0846	1.186559	0.775787	0.223112	1.089293	–	–
		1.631621	0.501316	0.526413	1.277349	0.553263	1.207343
		3.047413	0.245317	0.806738	1.745684	0.869291	2.766721
		21.331247	0.100882	0.957387	4.618576	0.975613	9.278486
		1.287083	0.026615	1.015763	1.134497	1.006360	8.032454
70.0	0.1080	1.239771	0.745308	0.204199	1.113450	–	–
		1.756106	0.502803	0.496069	1.325182	0.539196	1.217144
		3.370452	0.260186	0.784564	1.835879	0.854915	2.685183
		24.118420	0.112310	0.950399	4.911051	0.971379	8.354457
		1.357931	0.030508	1.017834	1.165303	1.007267	8.328375
75.0	0.1349	1.302412	0.718543	0.188462	1.141233	–	–
		1.897286	0.502883	0.469415	1.377420	0.526195	1.225104
		3.731835	0.273339	0.763826	1.931796	0.841394	2.615052
		27.221681	0.123187	0.943537	5.217440	0.967164	7.629533
		1.439962	0.034336	1.019848	1.199984	1.008167	8.623232
80.0	0.1657	1.375420	0.695021	0.175275	1.172783	–	–
		2.057025	0.501993	0.446040	1.434223	0.514268	1.231539
		4.136284	0.284927	0.744615	2.033786	0.828789	2.554375
		30.681457	0.133436	0.936896	5.539085	0.963034	7.051346
		1.534322	0.038053	1.021784	1.238677	1.009048	8.912666
85.0	0.2003	1.459843	0.674328	0.164157	1.208239	–	–
		2.237439	0.500458	0.425557	1.495807	0.503389	1.236722
		4.589115	0.295104	0.726951	2.142222	0.817119	2.501631
		34.543105	0.143012	0.930544	5.877338	0.959039	6.583655
		1.642303	0.041621	1.023627	1.281524	1.009901	9.193339
90.0	0.2389	1.556857	0.656106	0.154729	1.247741	–	–
		2.440918	0.498520	0.407609	1.562344	0.493511	1.240883
		5.096299	0.304026	0.710806	2.257498	0.806379	2.455610
		38.857483	0.151898	0.924531	6.233577	0.955215	6.200744
		1.765369	0.045013	1.025367	1.328672	1.010718	9.462762
95.0	0.2817	1.667791	0.640044	0.146694	1.291430	–	–
		2.670158	0.496353	0.391877	1.634062	0.484572	1.244217
		5.664537	0.311839	0.696119	2.380029	0.796540	2.415333
		43.681531	0.160096	0.918884	6.609200	0.951589	5.883966
		1.905172	0.048214	1.026997	1.380280	1.011495	9.719157
100.0	0.3288	1.794150	0.625872	0.139814	1.339459	–	–
		2.928205	0.494082	0.378079	1.711200	0.476504	1.246883
		6.301365	0.318677	0.682810	2.510252	0.787564	2.379994
		49.079189	0.167621	0.913620	7.005654	0.948179	5.619522
		2.063588	0.051212	1.028516	1.436519	1.012227	9.961344

N= 10   PRW= 0.5 DB

MSL	TW	A	B	C	WZ	WM	KM
50.0	0.0196	1.043333	0.646554	0.212489	1.021437	–	–
		1.978581	0.325205	0.613471	1.406621	0.718894	1.791547
		11.932366	0.115623	0.874558	3.454326	0.931032	7.514622
		1.242344	0.035648	0.972623	1.114605	0.983574	6.072132
		1.083350	0.007805	1.001499	1.040841	1.000362	9.735444
55.0	0.0292	1.115569	0.612176	0.187150	1.056205	–	–
		2.175423	0.334249	0.569140	1.474932	0.690580	1.787692
		13.696268	0.130856	0.847938	3.700847	0.915561	6.622772
		1.309914	0.043864	0.964619	1.144515	0.978928	5.965832
		1.064265	0.010145	1.001858	1.031632	1.000078	5.866197

ELLIPTIC LOW-PASS FILTER DATA 227

N= 10   PRW= 0.5 DB (CONTINUED)

MSL	TW	A	B	C	WZ	WM	KM
60.0	0.0416	1.155125	0.581840	0.166665	1.074768	–	–
		2.395586	0.340184	0.529495	1.547768	0.664472	1.780628
		15.646193	0.144838	0.821588	3.955527	0.899969	5.952956
		1.388319	0.052241	0.956131	1.178269	0.974039	5.887683
		1.091198	0.012663	1.002226	1.044604	1.000171	6.516847
65.0	0.0569	1.202764	0.555113	0.149943	1.096706	–	–
		2.641694	0.343772	0.494335	1.625329	0.640586	1.771620
		17.805157	0.157495	0.796069	4.219616	0.884599	5.437156
		1.478581	0.060589	0.947389	1.215969	0.969038	5.829401
		1.124860	0.015294	1.002596	1.060594	1.000281	7.178508
70.0	0.0754	1.259284	0.531580	0.136176	1.122178	–	–
		1.581882	0.345625	0.463327	1.257729	0.599711	1.544662
		20.199648	0.168834	0.771771	4.494402	0.869711	5.031667
		2.916792	0.068754	0.938598	1.707862	0.966435	9.579538
		1.165983	0.017978	1.002959	1.079807	1.000406	7.843975
75.0	0.0973	1.325557	0.510860	0.124753	1.151328	–	–
		1.699579	0.346226	0.436081	1.303679	0.582547	1.560265
		22.859830	0.178912	0.748947	4.781195	0.855494	4.707359
		3.224370	0.076619	0.929925	1.795653	0.961570	8.979299
		1.215334	0.020662	1.003313	1.102422	1.000540	8.506352
80.0	0.1227	1.402552	0.492612	0.115209	1.184294	–	–
		1.833223	0.345944	0.412193	1.353966	0.566975	1.572867
		25.819815	0.187818	0.727733	5.081320	0.842072	4.444196
		3.568394	0.084103	0.921500	1.889019	0.956822	8.489834
		1.273731	0.023304	1.003651	1.128597	1.000682	9.159270
85.0	0.1517	1.491357	0.476532	0.107186	1.221211	–	–
		1.984581	0.345060	0.391271	1.408751	0.552898	1.583010
		29.118013	0.195658	0.708182	5.396111	0.829518	4.228015
		3.953345	0.091151	0.913422	1.988302	0.952250	8.086046
		1.342073	0.025868	1.003972	1.158479	1.000826	9.797053
90.0	0.1846	1.593197	0.462354	0.100401	1.262219	–	–
		2.155654	0.343783	0.372957	1.468214	0.540210	1.591152
		4.384271	0.202539	0.690283	2.093865	0.813795	3.505075
		32.797559	0.097734	0.905759	5.726915	0.949059	9.482853
		1.421354	0.028331	1.004274	1.192206	1.000971	10.414826
95.0	0.2215	1.709458	0.449845	0.094635	1.307463	–	–
		2.348708	0.342266	0.356921	1.532549	0.528799	1.597672
		4.866846	0.208567	0.673983	2.206093	0.803361	3.439322
		36.906822	0.103839	0.898554	6.075098	0.944931	8.921289
		1.512694	0.030671	1.004556	1.229916	1.001113	11.008575
100.0	0.2624	1.841708	0.438800	0.089712	1.357095	–	–
		2.566301	0.340623	0.342875	1.601968	0.518556	1.602886
		5.407444	0.213844	0.659202	2.325391	0.793807	3.381585
		41.500011	0.109467	0.891830	6.442050	0.941050	8.457029
		1.617352	0.032878	1.004817	1.271752	1.001252	11.575156

N= 10   PRW= 1.0 DB

MSL	TW	A	B	C	WZ	WM	KM
55.0	0.0226	1.093729	0.505368	0.159171	1.045815	0.092490	1.001417
		2.044294	0.266577	0.566947	1.429788	0.711069	2.141872
		12.523802	0.099329	0.851980	3.538898	0.919962	8.676982
		1.264588	0.031715	0.965170	1.124539	0.980526	7.387325
		1.049952	0.007094	0.999696	1.024672	0.999335	6.816372

APPENDIX C

N= 10   PRW= 1.0 DB (CONTINUED)

MSL	TW	A	B	C	WZ	WM	KM
60.0	0.0332	1.128408	0.479555	0.141121	1.062265	0.093203	1.001870
		2.248946	0.272288	0.526273	1.499648	0.683967	2.135291
		14.349762	0.111075	0.824890	3.788108	0.904418	7.724906
		1.335806	0.038412	0.956268	1.155771	0.975614	7.284063
		1.072881	0.009051	0.999553	1.035800	0.999198	7.611045
65.0	0.0465	1.170678	0.456764	0.126417	1.081979	0.092798	1.002301
		2.477780	0.275868	0.490132	1.574097	0.659076	2.125462
		16.369340	0.121772	0.798480	4.045904	0.888993	6.999931
		1.418196	0.045162	0.947014	1.190880	0.970518	7.207548
		1.102063	0.011127	0.999391	1.049792	0.999062	8.424038
70.0	0.0629	1.221300	0.436665	0.114335	1.105125	0.091782	1.002701
		2.733566	0.277859	0.458221	1.653350	0.636375	2.113742
		18.606733	0.131397	0.773206	4.313552	0.873973	6.435236
		1.512833	0.051825	0.937636	1.229973	0.965364	7.150124
		1.138226	0.013272	0.999214	1.066877	0.998930	9.246453
75.0	0.0826	1.281096	0.418948	0.104330	1.131855	0.090449	1.003068
		3.019500	0.278690	0.430161	1.737671	0.615783	2.101087
		21.089677	0.139980	0.749371	4.592350	0.859568	5.987034
		1.620959	0.058291	0.928328	1.273169	0.960255	7.106569
		1.192112	0.015438	0.999028	1.087250	0.998804	10.069578
80.0	0.1057	1.350969	0.403330	0.095988	1.162312	0.088972	1.003401
		3.339232	0.278694	0.405552	1.827357	0.597183	2.088159
		23.849669	0.147583	0.727151	4.883612	0.845920	5.625648
		1.743996	0.064480	0.919244	1.320605	0.955274	7.073244
		1.234501	0.017588	0.998837	1.111081	0.998685	10.885191
85.0	0.1323	1.431927	0.389558	0.088987	1.196632	0.087458	1.003700
		3.696903	0.278125	0.384001	1.922733	0.580438	2.075403
		26.922266	0.154285	0.706624	5.188667	0.833118	5.330366
		1.883567	0.070337	0.910500	1.372431	0.950481	7.047563
		1.296237	0.019690	0.998644	1.138524	0.998574	11.685808
90.0	0.1627	1.525102	0.377409	0.083078	1.234950	0.085969	1.003969
		2.041516	0.277176	0.365140	1.428816	0.557878	1.889816
		30.347463	0.160175	0.687798	5.508853	0.821207	5.086361
		4.097182	0.075829	0.902179	2.024150	0.947462	9.786847
		1.368251	0.021718	0.998454	1.169723	0.998471	12.464857
95.0	0.1969	1.631774	0.366686	0.078064	1.277409	0.084544	1.004209
		2.219935	0.275987	0.348633	1.489945	0.545344	1.899048
		34.170146	0.165338	0.670633	5.845524	0.810199	4.882778
		4.545328	0.080939	0.894336	2.131977	0.943112	9.404157
		1.451578	0.023656	0.998268	1.204814	0.998377	13.216793
100.0	0.2352	1.753388	0.357216	0.073790	1.324155	0.083202	1.004422
		2.421187	0.274662	0.334180	1.556016	0.534102	1.906433
		38.440652	0.169858	0.655051	6.200053	0.800082	4.711513
		5.047250	0.085663	0.887003	2.246609	0.939028	9.081258
		1.547382	0.025490	0.998089	1.243938	0.998291	13.937139

N= 10   PRW= 2.0 DB

MSL	TW	A	B	C	WZ	WM	KM
65.0	0.0368	1.139941	0.356001	0.108561	1.067680	0.190507	1.060511
		2.313194	0.208130	0.493617	1.520919	0.678675	2.725271
		14.918842	0.087688	0.806102	3.862492	0.895444	9.700028
		1.358680	0.031094	0.949111	1.165624	0.972826	9.482897
		1.080729	0.007437	0.997452	1.039581	0.998380	10.392010

ELLIPTIC LOW-PASS FILTER DATA

N= 10   PRW= 2.0 DB (CONTINUED)

MSL	TW	A	B	C	WZ	WM	KM
70.0	0.0510	1.184571	0.339657	0.097691	1.088380	0.181846	1.062321
		2.549599	0.210202	0.460209	1.596746	0.654765	2.711418
		16.999407	0.095448	0.779821	4.123034	0.880244	8.842955
		1.444530	0.036200	0.939307	1.201886	0.967584	9.400138
		1.111876	0.009037	0.996865	1.054455	0.998056	11.467094
75.0	0.0683	1.237785	0.325232	0.088715	1.112558	0.174205	1.063881
		2.813843	0.211234	0.430795	1.677451	0.633009	2.695569
		19.305519	0.102401	0.754885	4.393805	0.865573	8.169553
		1.542973	0.041203	0.929485	1.242165	0.962324	9.337748
		1.150217	0.010675	0.996259	1.072482	0.997730	12.550096
80.0	0.0890	1.300428	0.312504	0.081251	1.140363	0.167463	1.065230
		3.109255	0.211514	0.404980	1.763308	0.613310	2.678827
		21.865983	0.108580	0.731528	4.676108	0.851603	7.631157
		1.655300	0.046030	0.919827	1.286585	0.957143	9.290256
		1.196504	0.012317	0.995646	1.093848	0.997406	13.629826
85.0	0.1131	1.373431	0.301274	0.075003	1.171935	0.161513	1.066398
		3.439622	0.211266	0.382366	1.854622	0.595541	2.661947
		24.713451	0.114040	0.709872	4.971262	0.838443	7.194354
		1.782991	0.050628	0.910474	1.335287	0.952115	9.253812
		1.251534	0.013937	0.995037	1.118720	0.997089	14.695847
90.0	0.1408	1.457835	0.291363	0.069742	1.207408	0.156258	1.067413
		3.809234	0.210659	0.362575	1.951726	0.579560	2.645434
		27.884732	0.118846	0.689954	5.280600	0.826157	6.835561
		1.927730	0.054961	0.901531	1.388427	0.947298	9.225662
		1.316173	0.015512	0.994442	1.147246	0.996784	15.738776
95.0	0.1723	1.554815	0.282613	0.065289	1.246922	0.151615	1.068295
		4.222929	0.209817	0.345260	2.054977	0.565221	2.629611
		31.421201	0.123063	0.671751	5.605462	0.814770	6.537726
		2.091433	0.059011	0.893067	1.446179	0.942729	9.203799
		1.391380	0.017026	0.993867	1.179568	0.996493	16.750483
100.0	0.2077	1.665695	0.274883	0.061500	1.290618	0.147510	1.069063
		4.686146	0.208836	0.330107	2.164751	0.552377	2.614676
		35.369275	0.126758	0.655201	5.947207	0.804280	6.288257
		2.276266	0.062767	0.885126	1.508730	0.938433	9.186744
		1.478226	0.018466	0.993318	1.215823	0.996218	17.724209

N= 10   PRW= 3.0 DB

MSL	TW	A	B	C	WZ	WM	KM
70.0	0.0441	1.403398	0.282052	0.090775	1.184651	0.216399	1.166668
		16.012345	0.170915	0.465732	4.001543	0.671018	3.911285
		2.437171	0.075326	0.786763	1.561144	0.883873	8.001590
		1.162946	0.027771	0.941675	1.078400	0.968511	6.709899
		1.096645	0.006807	0.996094	1.047208	0.997803	13.478362
75.0	0.0599	1.495877	0.269698	0.082151	1.223060	0.206334	1.168793
		18.210959	0.172036	0.435150	4.267430	0.647797	3.778244
		2.688174	0.081253	0.761130	1.639565	0.869040	7.723778
		1.212096	0.031893	0.931626	1.100952	0.963195	7.058382
		1.131573	0.008133	0.995302	1.063754	0.997390	14.802602
80.0	0.0790	1.601621	0.258792	0.074994	1.265552	0.197501	1.170494
		2.968753	0.172463	0.408289	1.723007	0.623529	3.246599
		20.650165	0.086537	0.737022	4.544245	0.856155	9.580625
		1.270276	0.035895	0.921683	1.127065	0.957935	7.390232
		1.174095	0.009474	0.994497	1.083557	0.996973	16.128511

N= 10   PRW= 3.0 DB (CONTINUED)

MSL	TW	A	B	C	WZ	WM	KM
85.0	0.1015	1.722022	0.249165	0.069015	1.312258	0.189740	1.171853
		3.282479	0.172396	0.384750	1.811761	0.605012	3.225969
		23.360803	0.091215	0.714599	4.833301	0.842720	8.998882
		1.338373	0.039725	0.912005	1.156881	0.952811	7.704447
		1.224986	0.010805	0.993695	1.106791	0.996562	17.442892
90.0	0.1275	1.858666	0.240667	0.063991	1.363329	0.182915	1.172938
		3.633404	0.171989	0.364147	1.906149	0.588340	3.205544
		26.377717	0.095338	0.693925	5.135924	0.830142	8.523227
		1.417375	0.043350	0.902713	1.190536	0.947884	8.000377
		1.285075	0.012107	0.992908	1.133611	0.996161	18.733688
95.0	0.1573	2.013360	0.233164	0.059745	1.418929	0.176906	1.173803
		4.026107	0.171356	0.346123	2.006516	0.573366	3.185808
		29.740120	0.098959	0.674995	5.453450	0.818457	8.129922
		1.508394	0.046748	0.893889	1.228167	0.943199	8.277696
		1.355276	0.013363	0.992145	1.164163	0.995775	19.990278
100.0	0.1908	2.188150	0.226535	0.056139	1.479240	0.171611	1.174493
		4.465741	0.170582	0.330354	2.113230	0.559946	3.167063
		33.492035	0.102134	0.657757	5.787230	0.807672	7.801582
		1.612680	0.049909	0.885587	1.269914	0.938784	8.536369
		1.436607	0.014564	0.991414	1.198585	0.995408	21.203666

APPENDIX D

TRANSITION WIDTHS (TW) FOR ELLIPTIC FILTERS

PRW= 0.1 DB

MSL	2	3	4	5	N 6	7	8	9	10
30.0	6.2301	1.4550	0.5304	0.2258	0.1025	0.0479	0.0227	0.0108	
35.0	8.6229	1.9331	0.7122	0.3129	0.1483	0.0728	0.0364	0.0183	
40.0		2.5195	0.9298	0.4176	0.2045	0.1045	0.0545	0.0288	0.0153
45.0		3.2359	1.1875	0.5414	0.2720	0.1434	0.0776	0.0425	0.0235
50.0		4.1088	1.4907	0.6858	0.3513	0.1901	0.1059	0.0600	0.0343
55.0		5.1703	1.8457	0.8527	0.4433	0.2449	0.1399	0.0814	0.0479
60.0		6.4597	2.2597	1.0444	0.5487	0.3082	0.1797	0.1070	0.0646
65.0		8.0246	2.7415	1.2632	0.6685	0.3805	0.2256	0.1371	0.0846
70.0		9.9228	3.3009	1.5122	0.8038	0.4622	0.2780	0.1718	0.1080
75.0			3.9495	1.7946	0.9558	0.5539	0.3371	0.2113	0.1349
80.0			4.7008	2.1140	1.1258	0.6563	0.4032	0.2558	0.1657
85.0			5.5704	2.4749	1.3156	0.7700	0.4767	0.3056	0.2003
90.0			6.5763	2.8818	1.5267	0.8957	0.5579	0.3607	0.2389
95.0			7.7394	3.3403	1.7611	1.0342	0.6473	0.4215	0.2817
100.0			9.0838	3.8564	2.0210	1.1866	0.7452	0.4882	0.3288

APPENDIX D

PRW= 0.5 DB

MSL	2	3	4	5	N 6	7	8	9	10
30.0	3.8087	0.9232	0.3244	0.1291	0.0539	0.0230	0.0099		
35.0	5.3829	1.2753	0.4611	0.1929	0.0857	0.0391	0.0180		
40.0	7.4892	1.7115	0.6284	0.2726	0.1270	0.0611	0.0299	0.0147	
45.0		2.2479	0.8298	0.3695	0.1785	0.0897	0.0460	0.0238	
50.0		2.9043	1.0692	0.4847	0.2410	0.1254	0.0668	0.0361	0.0196
55.0		3.7049	1.3518	0.6198	0.3151	0.1687	0.0928	0.0519	0.0292
60.0		4.6793	1.6832	0.7766	0.4014	0.2198	0.1243	0.0715	0.0416
65.0		5.8635	2.0704	0.9572	0.5008	0.2793	0.1615	0.0953	0.0569
70.0		7.3012	2.5213	1.1638	0.6141	0.3476	0.2047	0.1234	0.0754
75.0		9.0454	3.0454	1.3992	0.7425	0.4252	0.2542	0.1560	0.0973
80.0			3.6533	1.6664	0.8870	0.5125	0.3104	0.1934	0.1227
85.0			4.3578	1.9692	1.0490	0.6101	0.3734	0.2357	0.1517
90.0			5.1735	2.3113	1.2299	0.7187	0.4436	0.2831	0.1846
95.0			6.1173	2.6974	1.4314	0.8390	0.5213	0.3359	0.2215
100.0			7.2087	3.1326	1.6553	0.9718	0.6071	0.3941	0.2624

PRW= 1.0 DB

MSL	2	3	4	5	N 6	7	8	9	10
30.0	3.0041	0.7325	0.2504	0.0955	0.0380	0.0154			
35.0	4.3034	1.0366	0.3686	0.1495	0.0639	0.0279	0.0123		
40.0	6.0448	1.4162	0.5155	0.2187	0.0989	0.0460	0.0217		
45.0	8.3738	1.8851	0.6941	0.3042	0.1436	0.0702	0.0349	0.0175	
50.0		2.4606	0.9082	0.4072	0.1989	0.1013	0.0526	0.0277	
55.0		3.1640	1.1620	0.5292	0.2653	0.1395	0.0752	0.0411	0.0226
60.0		4.0212	1.4608	0.6716	0.3435	0.1855	0.1031	0.0582	0.0332
65.0		5.0639	1.8107	0.8364	0.4343	0.2395	0.1365	0.0793	0.0465
70.0		6.3305	2.2190	1.0257	0.5384	0.3020	0.1758	0.1045	0.0629
75.0		7.8679	2.6941	1.2419	0.6569	0.3734	0.2211	0.1342	0.0826
80.0		9.7327	3.2459	1.4880	0.7907	0.4543	0.2729	0.1684	0.1057
85.0			3.8858	1.7671	0.9411	0.5451	0.3314	0.2075	0.1323
90.0			4.6271	2.0830	1.1094	0.6464	0.3969	0.2515	0.1627
95.0			5.4851	2.4399	1.2973	0.7590	0.4696	0.3008	0.1969
100.0			6.4776	2.8424	1.5063	0.8836	0.5501	0.3554	0.2352

TRANSITION WIDTHS FOR ELLIPTIC FILTERS

PRW= 2.0 DB

MSL	2	3	4	5	N 6	7	8	9	10
30.0	2.2921	0.5568	0.1828	0.0659	0.0246				
35.0	3.3454	0.8144	0.2821	0.1098	0.0447	0.0185			
40.0	4.7610	1.1392	0.4084	0.1681	0.0732	0.0326			
45.0	6.6570	1.5433	0.5643	0.2419	0.1109	0.0524	0.0251		
50.0	9.1918	2.0414	0.7528	0.3324	0.1587	0.0786	0.0396		
55.0		2.6519	0.9780	0.4408	0.2171	0.1117	0.0587	0.0312	
60.0		3.3973	1.2444	0.5686	0.2869	0.1521	0.0828	0.0457	
65.0		4.3051	1.5574	0.7174	0.3687	0.2004	0.1123	0.0639	0.0368
70.0		5.4089	1.9236	0.8891	0.4633	0.2569	0.1474	0.0862	0.0510
75.0		6.7494	2.3505	1.0859	0.5715	0.3219	0.1884	0.1127	0.0683
80.0		8.3761	2.8470	1.3106	0.6943	0.3961	0.2356	0.1437	0.0890
85.0			3.4232	1.5660	0.8328	0.4798	0.2893	0.1793	0.1131
90.0			4.0913	1.8555	0.9883	0.5736	0.3498	0.2198	0.1408
95.0			4.8650	2.1829	1.1622	0.6781	0.4173	0.2654	0.1723
100.0			5.7603	2.5525	1.3561	0.7941	0.4923	0.3162	0.2077

PRW= 3.0 DB

MSL	2	3	4	5	N 6	7	8	9	10
30.0	1.9032	0.4581	0.1454	0.0502	0.0178				
35.0	2.8202	0.6878	0.2331	0.0878	0.0344	0.0137			
40.0	4.0558	0.9802	0.3466	0.1393	0.0589	0.0254			
45.0	5.7129	1.3460	0.4884	0.2058	0.0923	0.0425			
50.0	7.9300	1.7986	0.6615	0.2885	0.1353	0.0657	0.0324		
55.0		2.3547	0.8692	0.3885	0.1888	0.0955	0.0493		
60.0		3.0347	1.1160	0.5071	0.2533	0.1325	0.0710	0.0386	
65.0		3.8638	1.4067	0.6459	0.3294	0.1771	0.0980	0.0550	
70.0		4.8725	1.7475	0.8068	0.4180	0.2297	0.1304	0.0754	0.0441
75.0		6.0981	2.1453	0.9918	0.5198	0.2908	0.1687	0.0999	0.0599
80.0		7.5859	2.6085	1.2032	0.6357	0.3607	0.2130	0.1288	0.0790
85.0		9.3907	3.1465	1.4440	0.7668	0.4399	0.2637	0.1623	0.1015
90.0			3.7706	1.7173	0.9143	0.5290	0.3210	0.2005	0.1275
95.0			4.4936	2.0267	1.0796	0.6285	0.3852	0.2437	0.1573
100.0			5.3307	2.3763	1.2640	0.7391	0.4568	0.2921	0.1908

APPENDIX E

BESSEL LOW-PASS FILTER DATA

N=2

B	C
3.000000	3.000000

N=3

B	C
-	2.322185
3.677815	6.459433

N=4

B	C
5.792421	9.140131
4.207579	11.487800

N=5

B	C
-	3.646739
6.703913	14.272481
4.649349	18.156315

N=6

B	C
5.031864	26.514025
8.496719	18.801131
7.471417	20.852823

REFERENCES

[1] ANDAY, F., "Alternate state-variable realizations using single-ended operational amplifiers," *Proc. IEEE*, vol. 59, pp. 1710–1711, December 1971.

[2] BRANDT, R., "Active resonators save steps in designing active filters," *Electronics*, April 24, 1972, pp. 106–110.

[3] BRIDGMAN, A., and R. BRENNAN, "Simulation of transfer function using only one operational amplifier," *Proc. Wescon Convention Record*, vol. 1, pt. 4, 273–278 (1957).

[4] BUDAK, A., *Passive and Active Network Analysis and Synthesis*, Houghton Mifflin Company, Boston, 1974.

[5] DANIELS, R. W., *Approximation Methods for Electronic Filter Design*, McGraw-Hill Book Company, New York, 1974.

[6] DARYANANI, G., *Principles of Active Network Synthesis and Design*, John Wiley & Sons, Inc., New York, 1976.

[7] DELIYANNIS, T., "*RC* active allpass sections," *Electron. Letters*, vol. 5, pp. 59–60, February 1969.

[8] FLEISCHER, P. E., and J. TOW, "Design formulas for biquad active filters using three operational amplifiers," *Proc. IEEE*, vol. 61, no. 5, pp. 662–663, May 1973.

[9] GRAEME, J. G., G. E. TOBEY, and L. P. HUELSMAN (eds.), *Operational Amplifiers: Design and Applications*, McGraw-Hill Book Company, New York, 1971.

[10] GREBENE, A. B., *Analog Integrated Circuit Design*, Van Nostrand Reinhold, New York, 1972.

[11] GUILLEMIN, E. A., *Synthesis of Passive Networks*, John Wiley & Sons, Inc., New York, 1957.

[12] HILBURN, J. L., and D. E. JOHNSON, *Manual of Active Filter Design*, McGraw-Hill Book Company, New York, 1973.

[13] HUELSMAN, L. P., *Theory and Design of Active RC Circuits*, McGraw-Hill Book Company, New York, 1968.

[14] HUELSMAN, L. P., *Active Filters: Lumped, Distributive, Integrated, Digital, and Parametric*, McGraw-Hill Book Company, New York, 1970.

[15] INIGO, R. M., "Active filter realization using finite-gain voltage amplifiers," *IEEE Trans. Circuit Theory*, vol. CT-17, pp. 445–448, August 1970.

[16] JOHNSON, D. E., *Introduction to Filter Theory*, Prentice-Hall, Inc., Englewood Cliffs, N.J., 1976.

[17] JOHNSON, D. E., and J. L. HILBURN, *Rapid Practical Designs of Active Filters*, John Wiley & Sons, Inc., New York, 1975.

[18] KERWIN, W. J., and L. P. HUELSMAN, "The design of high performance active RC band-pass filters," *IEEE International Convention Record*, vol. 14, pt. 10, pp. 74–80, 1960.

[19] LINDQUIST, C. S., *Active Network Design*, Steward and Sons, Long Beach, Calif., 1977.

[20] MELEN, R., and H. GARLAND, *Understanding IC Operational Amplifiers*, Howard W. Sams and Co., New York, 1971.

[21] MITRA S. K. (ed.), *Active Inductorless Filters*, IEEE Press, New York, 1971.

[22] MITRA, S. K., *Analysis and Synthesis of Linear Active Networks*, John Wiley & Sons, Inc., New York, 1969.

[23] MOORE, H. P., D. E. JOHNSON, and J. R. JOHNSON, "Active elliptic filters," *Proc. 1976 IEEE Southeastern Conference and Exhibit*, pp. 335–336, April 1976.

[24] PAPOULIS, A., "On the approximation problem in filter design," *IRE National Convention Record*, vol. 5, pt. 2, pp. 175–185, 1957.

[25] RICHARDS, P., "Universal optimum-response curve for arbitrarily selected coupled resonators," *Proc. IRE*, vol. 34, pp. 624–629, September 1946.

[26] SALLEN, R. P., and E. L. KEY, "A practical method of designing RC active filters," *IRE Trans. Circuit Theory*, vol. CT-2, pp. 74–85, March 1955.

[27] STORCH, L., "Synthesis of constant-time-delay ladder networks using Bessel polynomials," *Proc. IRE*, vol. 42, no. 11, pp. 1666–1675, November 1954.

[28] SU, K. L., *Active Network Synthesis*, McGraw-Hill Book Company, New York, 1965.

[29] TEMES, G. C., and S. K. MITRA, *Modern Filter Theory and Design*, John Wiley & Sons, Inc., New York, 1973.

[30] THOMSON, W. E., "Delay networks having maximally flat frequency characteristics," *Proc. IEE*, vol. 96, pt. 3, pp. 485–490, November 1949.

[31] TOW, J., "Design formulas for active RC filters using operational amplifier biquad," *Electron. Letters*, pp. 339–341, July 24, 1969.

[32] VAN VALKENBURG, M. E., *Introduction to Modern Network Synthesis*, John Wiley & Sons, Inc., New York, 1960.

[33] WEINBERG, L., *Network Analysis and Synthesis*, McGraw-Hill Book Company, New York, 1962.

INDEX

INDEX

A

Active filter, 6
All-pass constant time delay, 166
   design procedures, 167, 173
All-pass filter, 4, 157, 166
   amplitude response of, 157
   biquad, 161, 170
   design procedures, 159, 161, 168, 170
   multiple feedback, 159, 168
All-pole approximation, 5
Amplitude response, 2
   of all-pass filter, 157
   of bandpass filter, 94
   of band-reject filter, 129
   of Bessel (constant-time delay) filter, 166
   in decibels, 3
   of high-pass filter, 70, 71
   of low-pass filter, 11
Anday, F., 239

B

Band-elimination filter, *see* Band-reject filter
Bandpass filter, 93
   amplitude response of, 94
   bandwidth of, 93
   biquad, 117, 125
   Butterworth, 94, 98
   center frequency of, 93
   Chebyshev, 94, 98
   cutoff points of, 93, 95
   design procedures, 103, 105, 106, 110, 111, 114, 117, 119, 123, 125
   elliptic, 94, 100, 107, 119
   gain of, 94
   ideal, 93
   infinite-gain, multiple-feedback, 103, 111
   inverse Chebyshev, 94, 100, 107, 119
   passband of, 93
   quality factor of, 94

242 INDEX

Bandpass filter *(cont):*
  stopband of, 93
  three capacitor elliptic, 123
  transfer function of, 98
  transition bands of, 93
  transition widths of, 101
  VCVS, 105, 114
  VCVS elliptic, 119
Band-reject filter, 129
  amplitude response of, 129
  bandwidth of, 129
  biquad, 153
  Butterworth, 133
  center frequency of, 129
  Chebyshev, 133
  cutoff points of, 129
  design procedures, 137, 138, 141, 144, 145, 149, 153
  elliptic, 134
  gain of, 130
  infinite-gain, multiple-feedback, 137
  inverse Chebyshev, 134
  passband of, 129
  quality factor of, 130
  stopband frequencies of, 131
  stopband of, 130
  three capacitor elliptic, 149
  transfer function of, 133
  transition bands of, 130
  transition widths of, 135
Bandwidth:
  of bandpass filter, 93
  of band-reject filter, 129
Bessel (constant-time delay) filter, 163
  amplitude response of, 166
  cutoff frequency of, 166
  design procedures, 172
  phase response of, 163
  time delay of, 163
  transfer function of, 164
Biquad filter:
  all-pass, 161
  bandpass, 106, 117, 125
  band-reject, 153
  high-pass, 75, 83, 88
  low-pass, 28, 36, 57, 65
  VCVS, 26, 34

C

Capacitors, types of, 8
Cascading, 8

Chebyshev filter, 13, 17
  amplitude response of, 17
  bandpass, 94, 98
  band-reject, 133
  biquad, 28, 36
  high-pass, 69
  infinite-gain, multiple-feedback, 22, 23
  low-pass, 17
  phase response of, 16
  ripple width of, 17
  VCVS, 26, 34
Chebyshev polynomials, 17, 41
Constant time-delay filter, *see* Bessel filter
Cutoff frequency, 2

D

Daniels, R. W., 239
Daryanani, G., 239
Deliyannis, T., 239
Design data:
  Butterworth, 177
  Chebyshev, 177
  elliptic, 191, 233
  inverse Chebyshev, 181
Design procedure, *see* Specific type of filter

E

Elliptic filter, 46
  amplitude response of, 46
  bandpass, 94, 100, 107, 119
  band-reject, 134
  biquad, 57, 88
  high-pass, 70
  low-pass, 46
  phase response of, 16
  ripple channel of, 46
  three capacitor, 54, 63, 87
  VCVS, 50, 61, 85

F

Filter:
  active, 6
  all-pass, *see* All-pass filter
  bandpass, *see* Bandpass filter
  band-reject, *see* Band-reject filter
  Bessel, *see* Bessel filter
  Butterworth, *see* Butterworth filter

INDEX

Filter *(cont.)*:
   cascading, 8
   Chebyshev, *see* Chebyshev filter
   constant-time delay, *see* Bessel filter
   construction of, 8
   electric, 1
   elliptic, *see* Elliptic filter
   frequency selective, 1
   high-pass, *see* High-pass filter
   inverse Chebyshev, *see* Inverse Chebyshev filter
   low-pass *see* Low-pass filter
   order of, 5
   passive, 6
   phase-shift, *see* All-pass filter
   realizable, 5
Fleisher, P. E., 240

G

Gain, 12
Garland, H., 240
Graeme, J. G., 240
Grebene, A. B., 240
Guillemin, E. A., 240

H

High-pass filter, 69
   amplitude response of, 70, 71
   biquad, 75, 83
   Butterworth, 69
   Chebyshev, 69
   cutoff points of, 69
   design procedures, 72, 74, 75, 76, 80, 81, 83, 85, 87, 88, 90
   elliptic, 70, 76, 85, 87, 88
   gain of, 70
   ideal, 69
   infinite-gain, multiple-feedback, 72, 80
   inverse Chebyshev, 70, 76, 85
   passband of, 69
   stopband of, 69
   transfer function of, 69, 70, 71, 76
   transition band of, 69, 71
   VCVS, 74, 81
Hilburn, J. L., 240
Huelsman, L. P., 240

I

Infinite-gain, multiple-feedback filter:
   bandpass, 103, 111
   band-reject, 137
   high-pass, 72, 80
   low-pass, 22, 33
Inigo, R. M., 240
Integrated circuit, *see* Operational amplifier
Inverse Chebyshev filter, 41
   amplitude response of, 41
   bandpass, 94, 100, 107, 119
   band-reject, 134
   cutoff point of, 41
   high-pass, 70
   low-pass, 41
   phase response of, 16
   ripple channel of, 42
   transfer function of, 44
   transition width of, 43

J

Johnson, D. E., 240
Johnson, J. R., 240

K

Kerwin, W. J., 240
Key, E. L., 241

L

Lindquist, C. S., 240
Linear-phase filter, 163
Low-pass filter, 11
   amplitude response of, 11
   biquad, 57, 65
   Butterworth, 13, 14
   Chebyshev, 13, 17
   cutoff frequency of, 2
   design procedures, 22, 26, 28, 31, 33 34, 36, 38, 50, 54, 61, 63, 65
   elliptic, 13, 41
   gain of, 12
   ideal, 2, 5
   infinite-gain, multiple-feedback, 22, 23
   optimum, 13
   passband of, 2, 11
   stopband of, 3, 11

Low-pass filter *(cont.):*
  transition band of, 2, 11
  VCVS, 50, 61

M

Melen, R., 240
Mitra, S. K., 240, 241
Moore, H. P., 240
Multiple-feedback filter, *see* Infinite-gain, multiple-feedback filter

N

Notch filter, *see* Band-reject filter

O

Odd-order filters, 31, 38, 60, 66, 79, 90
Op-amp, *see* Operational amplifier
Operational amplifier, 6, 7,
  manufacturers of, 7
  types of, 7
  slew rate of, 7
Order, 5
  minimum, 20, 42, 48, 72, 102, 136

P

Papoulis, A., 241
Passband, 2
Phase response, 2, 4
  of all-pass filter, 158
  of Bessel filter, 163
Phase-shift filter, *see* All-pass filter
Pole, 5
Pole-pair frequency, 9
Pole-pair quality factor, 9

Q

Q, *see* Quality factor
Quality factor:
  bandpass, 94
  band-reject, 130
  pole-pair, 9, 54, 71, 99

R

Resistors, types of, 7
Richards, P., 241
Ripple width, 17, 46

S

Sallen, R. P., 241
Stopband, 2
Storch, L., 241
Su, K. L., 241

T

Temes, G. C., 241
Thomson, W. E., 241
Three-capacitor elliptic filter, 54, 63, 87, 123, 149
Time delay, 4
Time-delay filter, *see* Bessel filter
Tobey, G. E., 240
Tow, J., 240, 241
Transfer function, 1, 5
  of all-pass filter, 158
  of bandpass filter, 98-101
  of band-reject filter, 133-135
  of Bessel filter, 164
  of high-pass filter, 69-71, 76
  of low-pass filter, 15, 31, 50
Transition band, 2
Transition width, 13, 43, 46, 71, 101, 136, 233
  normalized, 43
Tuning, 58, 77, 108, 140

V

Van Valkenburg, M. E., 241
Voltage-controlled-voltage source filter, *see* VCVS filter
VCVS filter, 50, 61, 85, 105, 114, 119, 144, 145

W

Weinberg, L., 241

Z

Zero, 5